Hospital Wastewater Treatment

Global Scenario and Case Studies

Hospital Wastewater Treatment

Global Scenario and Case Studies

Edited by

Nadeem A Khan, Viola Vambol,
Sergij Vambol, Nastaran Mozaffari and
Niloofar Mozaffari

IWA PUBLISHING

Published by **IWA Publishing**
Unit 104–105, Export Building
1 Clove Crescent
London E14 2BA, UK
Telephone: +44 (0)20 7654 5500
Fax: +44 (0)20 7654 5555
Email: publications@iwap.co.uk
Web: www.iwapublishing.com

First published 2022

British Library Cataloguing in Publication Data
A CIP catalogue record for this book is available from the British Library

ISBN: 9781789062618 (Paperback)
ISBN: 9781789062625 (eBook)
ISBN: 9781789062632 (ePub)

Contents

Chapter 3

Chapter 4

doi: 10.2166/9781789062625_xi

Preface

The remarkable vision of Dr. Viola Vambol, Dr. Sergi Vambol, Dr. Mika Sillanpää, Dr. (Prof.) Mohammad Hadi Dehghani, Dr. (Prof.) Sirajuddin Ahmed, JMI, Dr. (Prof.) Izharul Haq Farooqi and Dr. (Prof.) Imran Ali, JMI motivated all the authors involved in the production of this book. Their remarkable and in-depth knowledge of Environmental Engineering has considerably helped the authors during this work.

In recent years, the treatment of emerging contaminants has emerged as a new field, drawing the attention of scientists working towards describing emerging contaminants entirely in terms of pathways and innovative treatment methods. Topics include sources, occurrences, environmental management, sustainable development, and innovative treatment methods. These topics are of great importance due to their impact on biotic systems, innovation, and development. This book will provide an advanced overview of the actual situation regarding hospital wastewater in different countries. In public sewage systems and surface water, which acts as an endpoint, hospital effluents are deemed a point source of micropollutants such as medication residues (antibiotic resistance genes) and antibiotic-resistant bacteria. It is pivotal to start with research into the fate and mechanisms of medicinal substances and their metabolites.

Pharmaceutical contamination is increasingly recognized as a hazard to habitats and human health worldwide, and it may also lead to antimicrobial-resistance growth. Although hospitals have a special therapeutic mission, chemical contaminants in hospital wastewater are a significant source of contamination with possible environmental and health threats. Between 30 and 90% of orally ingested pharmaceuticals are excreted in wastewater as active substances in the feces and urine of patients. The ability of urban wastewater treatment plants to eliminate these active contaminants varies, which means they may be released into the marine system and thereby join the water cycle. This volume will address the needs of researchers as well as graduate students, whilst also helping policymakers to understand the current scenario and implement appropriate treatment systems. This book will help understand the case studies around the globe to deal with hospital wastewater effectively. All chapters are published collectively here under the title 'Hospital Wastewater Treatment: Global scenario and case studies'. The chapters in this book cover new topics, and any suggestions for further research are always welcomed.

Editors
Nadeem A Khan
Viola Vambol
Sergij Vambol
Nastaran Mozaffari
Niloofar Mozaffari

doi: 10.2166/9781789062625_0001

Chapter 1

Health risk associated with hospital effluents

Aarfa Queen[1], Mohammad Yusuf[2], Abdul Qadir[3], and Sakina Aamir[4]*

[1]Department of Chemistry, Jamia Millia Islamia, New Delhi 110025, India
[2]Department of Chemical Engineering, Centre of Contamination Control and Utilization (CenCoU), Universiti Teknologi PETRONAS, Bandar Seri Iskandar, 32610, Perak, Malaysia
[3]Department of Architecture, Faculty of Architecture and Ekistics, Jamia Millia Islamia, New Delhi 110025, India
[4]Department of Genetics, University of Delhi South campus, New Delhi 110025, India
**Corresponding author. yusufshaikh.amu@gmail.com*

ABSTRACT

Hospital wastewater is a major biohazard and a leading cause of the spread of multiple diseases. The outbreak of the COVID-19 pandemic further imposes a significant threat to the monitoring and management of various infectious pathogens, including viruses, bacteria and other microbes present in hospital wastewater. These effluents consist of potential chemicals such as organic pollutants and pharmaceutically active compounds (PhACs), which are hazardous to human health. The major effluent contributors come from multiple hospital activities including drug treatments, disinfecting, application of detergents, surgery, premises cleaning, and chemical or biological analysis in laboratories. Most of these contributors are recalcitrant as they persist in the receiving environment, leading to biomagnification through food chains and bioaccumulation in living organisms, including humans. Wastewaters produced from hospitals also contain radionuclides, chemical reagents, and other toxic heavy metals. In addition, biological contaminants like faecal indicator bacteria (FIB) and antibiotic-resistant genes (ARGs) in marine environments remain a global problem, resulting in potentially severe risks to animal and human health. Numerous reports relate to the introduction and spreading of contaminants into various components of the environment. ARGs, toxic metals and FIB released from hospital wastewater effluents build up in the sediments of the rivers receiving them. ARGs formed due to resistance caused by antibiotics, such as monobactams, penicillins, and cephalosporins, in beta-lactamases quantified through quantitative polymerase chain reaction (qPCR) of species-specific genes in complete DNA extract recovered from sediments of hospital discharges and river systems, show a high concentration of toxic metals such as Cr, Zn, Cu, Pb, Cd, and Hg. Multiple findings suggest that numerous sources, involving urban and hospital wastewaters, are contributing to dispersal of biological contaminants and toxic metals in aquatic ecosystems. Additionally, the bioactive intrinsic properties of various pharmaceuticals and drugs discarded into hospital effluents can also cause potential effects in aquatic organisms. These residual products, directly and indirectly, pose various health risks and contribute to development

Editor(s): Nadeem A Khan, Viola Vambol, Sergij Vambol, Nastaran Mozaffari and Niloofar Mozaffari

of deadly diseases such as shigelloses, cholera, lung infection, asthma, dysentery, and multiple types of cancers. This chapter provides detailed information on the occurrence, perseverance, associated health risk, preventive measures, and elimination of drug molecules, drug-resistant pathogens, including infectious viruses, other microorganisms, and micropollutants.

Keywords: Antibiotic resistance, Hospital wastewater, Recalcitrant, Sediment, Toxic metals, Water pollution

CONTENTS

1.1 INTRODUCTION

The increasing population of the world and its rising demands for natural resources are posing a burden on natural resources (Rosdin *et al.*, 2021; Yusuf *et al.*, 2021a, 2021b, 2021c, 2021d, 2021e). The exploitation of natural resources (i.e., mainly fossil fuels for energy and water demands) on this scale is not only altering the global climate, but also deteriorating the quality of the environment (Iqbal *et al.*, 2021; Yusuf *et al.*, 2020, 2021e;). Water is a fundamental commodity, that is, essential for human survival, and it is present in abundance. Almost 75% of the earth's surface is covered with water as oceans (Hamid *et al.*, 2021; Yusuf *et al.*, 2021d). However, all that water cannot be utilized directly for our day-to-day needs. The water that is utilized directly by us is the groundwater, that is, that which is present in the underground water tables, and it is very limited (Yusuf, 2016). The growing population, industrialization, and exploitation of water tables at an extreme level by humans has led to a decrease in water table levels during the last five decades (Alam *et al.*, 2021; Yusuf *et al.*, 2021b, 2021c). Recently, to confront these challenges, various governments of the world came together with stringent laws for water regulation, and harvesting and treatment of wastewaters (Vambol *et al.*, 2021b). For the treatment of wastewaters, their classification on the basis of type of contaminant present is of the key importance (Khan *et al.*, 2021a). There are various government regulations and treatment methods for industrial, domestic, nuclear, chemical laboratory and hospital wastewaters (Khan *et al.*, 2021b). This study mainly compiles the classifications and techniques for treatment of wastewater, obtained from pharmaceutical and hospital effluents.

With the increasing awareness of water pollution and its harmful human health and wellbeing effects, great attention is being given to addressing emerging pollutants in surface waters, wastewaters and groundwaters. Furthermore, the outbreak of the COVID-19 pandemic further reinforces the need for careful monitoring and management of wastewater in view of the various infectious pathogens which may be present. A significant cause of illness and disease is contamination of drinking water by bacteria and viruses from human and animal faecal matter and waterborne pathogens. Polluted and contaminated water is linked to the spread of diseases such as cholera, diarrhoea, dysentery, hepatitis A, typhoid, and polio. It is estimated that contaminated drinking water causes 485 000 diarrhoeal deaths each year. Unfortunately, most of these deaths are highly preventable if pollution of water bodies can be controlled or at least reduced (Laffite *et al.*, 2016). However, globally there is a lack of sufficient wastewater management and treatment, especially in developing countries. As per the UN records, more than 80% of the wastewater of the world is not treated and is released without being reused. Emerging contaminants are unregulated pollutants, which may require regulation in the future, depending on the emergence of research or new data on their health effects. They may include any kind of chemical compounds including surfactants, pharmaceutical and personal care products (PPCPs), endocrine disruptors, illegal drugs, and fuel additives (Joss *et al.*, 2006).

1.2 HOSPITAL WASTEWATER

Hospitals have become an integral part of every city, and they are important for ongoing human medical research. At the same time, they are also a source of prominent water pollutants (Vambol *et al.*, 2021a). In developed countries, 400–1200 L of wastewater is generated per bed per day in hospitals (Kumari *et al.*, 2020). Hospital activities that generate harmful pollutants include drug treatments, application of disinfectants and detergents, surgery, chemical or biological analysis in laboratories, diagnostics, research, and creation of human body waste such as excreta or samples from patients. Waste from hospitals is loaded with pharmaceutically active compounds, for example analgesics, antidepressants, anti-convulsants, antibiotics, anti-inflammatories and hormones, that have harmful effects on flora and fauna, and water bodies as well as on other animals and humans that depend on those water bodies (Kozub *et al.*, 2021). Hospital wastewater (HWW) is thus proving to be a significant biohazard and a leading cause of the spread of multiple diseases. HWW comprises heterogeneous mixtures of biological and chemical substances from

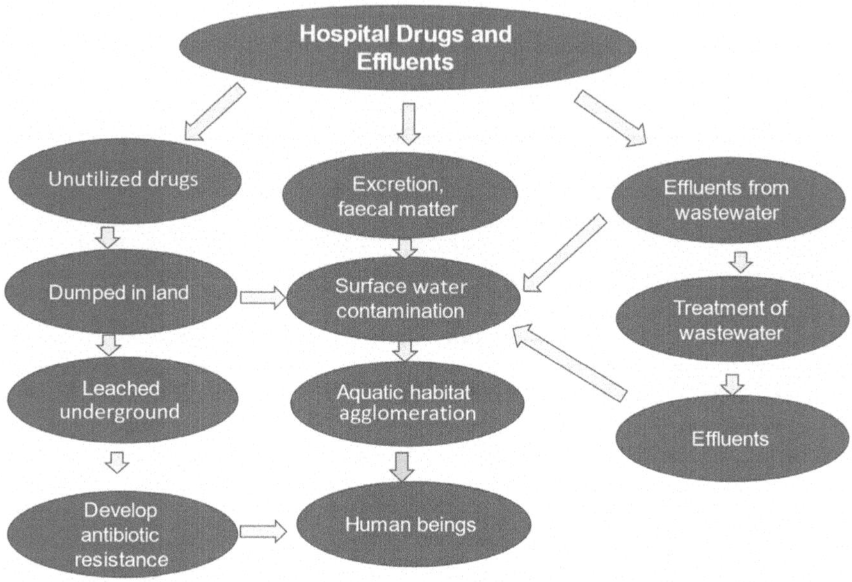

Figure 1.1 Hospital effluents and risk factors.

diagnostic laboratories; waste from research activities and medicine excretion, and their metabolites; chemicals; disinfecting agents; iodinated contrast media; radioactive markers; nutrients; and bacteria and their antimicrobial-resistant genes.

If discharged into cesspool systems, and other marine bodies (i.e., lakes, rivers, seas etc.) without preceding treatment, this hospital waste can accumulate in sediments over time, leading to very harmful long-lasting effects on the life of surrounding flora and fauna, including human health and longevity (Figure 1.1).

1.2.1 Classification of hospital wastewater

The categorization of hospital waste by the World Health Organization (WHO) is as follows (Zotesso *et al.*, 2017):

(1) Blackwater (sewage) comprising urine and faecal matter.
(2) Greywater (sullage) comprising waste and residuals from bathing, washing, laundry, lab processes, and a number of various technical processes like cooling of water, or the cleaning of X-ray films, containing cytotoxic and genotoxic or chemicals.
(3) Stormwater being rainwater collected from grounds, roofs, paved surfaces, yards, rinsing water for hospital grounds, flushing toilets, and other a number of washing purposes, which perhaps gets flushed to drains and groundwater recharge.

Further, domestic discharges are the discharges from domestic sources like laundries, kitchens, and toilets of regular wards. Specific discharges are releases of wastewater from activities of research laboratories and other chemicals (like dyes, acids, alkalis, solvents like benzene and other hydrocarbons) (US Environmental Protection Agency, 2015).

1.2.2 Contents of hospital wastewater

The waste generated from hospitals includes a variety of chemical compounds.

1.2.2.1 Drugs and their metabolites

Use and administration of various drugs across the world are continuously increasing. The human body partially metabolizes the active chemicals, and the remaining unmetabolized active substances are expelled, mainly through urine or faeces, thus entering the water cycle (Joss *et al.*, 2006).

1.2.2.2 Bacteria, fungi and viruses

Hospital waste includes excreta and faeces, especially from patients. Human faecal matter contains approximately 100 billion (1×10^{11}) bacteria. The bacteria include some specific pathogenic bacteria, like *Campylobacter* or *Salmonella*, causing gastroenteritis. Many of the known bacteria that reside in the intestinal tracts of humans and are excreted along with faeces are used as markers for measuring faecal levels of contamination of water. These are called faecal indicator bacteria and include *Enterococci* and *E. coli*. These bacteria can continue to multiply, far away from their habitat in the animal gut. If enough of these bacteria are ingested, these pathogens can cause disease. In addition, the spores of fungi are multiplied at a faster rate in the external environment (compared to in the animal gut) posing a threat to human health and the environment. The microorganisms that are pathogenic and are present in HWWs, enter the HWW from the atmosphere, medical devices, and water utilized in hospitals. Thus HWW comprises pathogenic microbes, like algae, bacteria, yeast, fungi, viruses, parasites, protozoa, and bacteriophages (US Environmental Protection Agency, 2015).

1.2.2.3 Antibiotics

One of the most harmful chemicals released from hospitals are antibiotics. The administration of antibiotics has been increasing over time, resulting in their accumulation in the environment. In a study by Ngigi *et al.* (2020) in Kenya, residues of antibiotics were 3–10 times higher in HWW as compared to those detected in a wastewater treatment plant (WWTP) and surface water. In this study, antibiotics such as trimethoprim used in the treatment of bladder infection, spectinomycin used in gonorrhoeal infection, and ampicillin as well as oxacillin used in a number of bacterial infections, were observed in the samples of wastewater from hospitals. As soon as the antibiotic residues are released into the environment, they affect biota at various trophic levels thereby affecting human health due to the development of increasing resistance in various bacterial populations that causes pressure in the associated sections of the environment. Due to the excessive pollution by antibiotics, there is increasing danger of the development of so-called 'super-bugs' – microbes that are highly resistant and unresponsive to available drugs – which pose a significant challenge to human health.

1.2.2.4 *Antibiotic-resistant bacteria and antibiotic-resistant genes*

Though antibiotics have led to a reduction in mortality due to bacterial diseases, their misuse at a higher level has resulted in exponential augmentation in the development of antibiotic and multidrug-resistant bacteria all over the world. HWW is the major source of antibiotic-resistant genes (ARGs) that emerge due to the continuous release of antibiotic pools from hospitals, increasing the population of resistant bacteria. These resistant bacteria either act as vectors for carrying transmissible genes or as the ARG reservoirs that pose a great danger to public health. Antibiotic resistance has become a global problem that results in serious health risks for human as well as animal health.

1.2.2.5 *Lipid regulators*

Lipid regulators along with their metabolites, clofibric acid and fenofibric acid, remain in water. Although there is limited information on their harmful effects, several studies have indicated the need for controlling these pollutants (Hernando *et al.*, 2007). Hormones, oestrogens, and endocrine-disrupting compounds (EDCs) – hormones such as 17β-oestradiol and 17α-ethinyloestradiol – are frequently detected in HWW (Schröder *et al.*, 2016). 17β-Oestradiol plays a primary role in female sex hormones and it is a critical regulator of reproductive cycles like the oestrous cycle in animals such as cows and buffalos, and the menstrual cycle in human females. 17α-Ethinyloestradiol is a bioactive synthetic oestrogen that is used in combined oral contraceptive pills. Both oestrogens, 17α-ethinyloestradiol and 17β-oestradiol, are found in nanogram-per-litre concentrations and trigger endocrine-disrupting effects in the ecological community. Thus EDCs cause decreased fertility in human beings, changes in sexual behaviour, and an increase in cancers and other malformations in human beings and other faunae (Gadupudi *et al.*, 2021).

1.2.2.6 *Heavy metals*

One of the important sources of toxic and noxious elements, for example gadolinium (Gd), mercury (Hg), platinum (Pt), and other heavy metals like Cu, Cd, Pb, Fe, Ni, and Zn, is hospital wastewater. HWW from dentistry departments is an important source of amalgam, and discharges mercury, silver (Ag), tin (Sn), copper (Cu), and zinc (Zn) to water bodies. It has been suggested that both ARGs and metals lie on similar mobile elements, which leads to collection of ARGs by these metals (Seiler & Berendonk, 2012).

1.2.2.7 *Disinfectants, detergents and sterilization products*

Detergents affect ecosystems by affecting the flora and fauna. Aerobic processes can lead to the degradation of most of the detergents, but anaerobic degradation does not take place due to restricted metabolic pathways and the toxicity linked with it (Mousavi & Khodadoost, 2019). The major components of detergents are phosphates. The presence of these phosphates in water bodies leads to eutrophication thereby resulting in excessive algal growth. These excessive algae cause an increase in the organic substances and hence a reduction in levels of dissolved oxygen in water, which hampers the life of aquatic flora and fauna. The main parameters of water, like salinity, temperature, turbidity, pH and so on., are altered by the phosphates and other chemical surfactants. Thus, wastewater having high detergent concentrations can lead to a reduction in efficiency of aerobic systems and a reduction in gas production in anaerobic systems and so these factors, including eutrophication, need to be controlled.

1.2.2.8 Cytostatic agents

Cytostatic drugs are those natural as well as synthetic compounds that are used in the treatment of cancer including chemotherapy. The main role of these anticancer drugs is in cell division, for blocking the replication of DNA in the tumour cell and causing cell death. These compounds also inhibit nucleotide synthesis and intercalate in DNA, thereby preventing transcription and translation. According to the EU Commission Decision 2000/532/EC251, that established the European List of Waste (LoW), the cytostatic and cytotoxic drugs are thought to be a hazardous waste because they (even at ultra-low doses) cause DNA mutations (Jureczko & Kalka, 2020). Hence, the accumulation of such chemicals in a water body is hazardous. Furthermore, it can cause mutations in the organisms inhabiting the water body or even in humans consuming the polluted water.

1.2.2.9 Radioactive markers

Radioactive waste can hamper the life of many generations of animals as well as plants as it can cause mutations in DNA and also cancerous growths. Radioactive isotopes are continuously in use for therapeutic and diagnostic applications in hospitals and trauma centres. Some important radioactive isotopes used in hospitals are carbon-14 (C-14), technetium-99m (Tc-99m), iodine-131 (I-131), iodine-125 (I-125), flourine-18 (F-18), iodine-123 (I-123), and tritium (H-3). The solid waste comprising needles, syringes, vials, cotton swabs, absorbent materials, contaminated gloves, and the clothes and utensils of radioactive-isotope-treated patients constitute radioactive solid waste. Also, the excreta and urine of patients treated with a high dose of radioisotopes contain traces of radioactive materials. The first and foremost important strategy of hospital waste management for such waste is its safe disposal. Exposure to radiation, for the public and radiation workers, from the environment should not exceed the safe prescribed limits, so that the short and long-term impacts of ionizing radiation can be reduced (Ravichandran, 2017). Radioactive waste can remain in the environment for decades, thereby making its disposal a great challenge. Also, the accidental release of these contaminants due to their improper disposal may disturb marine, surface water and groundwater resources.

1.2.2.10 Iodinated contrast media

Iodinated contrast media (ICM), such as iomeprol, are used for hospital diagnostics and imaging. Research has shown that these chemicals are harmful to the body. For example, Carmona-Sánchez *et al.* showed that iodinated intravenous contrast medium harms the human pancreas by reducing pancreatic blood flow (Carmona-Sánchez *et al.*, 2000). Many of the compounds mentioned above are emerging contaminants, which are often regarded as unregulated pollutants, and they require future regulation monitoring their occurrence and potential effects on health. The different types of hospital wastes are categorized in Figure 1.2.

1.2.3 Biomagnification

Most of the pollutants discussed above remain in the ecosystem for extended durations of time, leading to biomagnification through food chains and bioaccumulation in living organisms, including humans. Biomagnification is defined as the increase in the chemical concentration and bio-amplification of toxins in the tissues of organisms at successively higher trophic levels in a food chain. Harmful chemicals, such as drugs or heavy metals from hospital effluents, affect rivers and oceans and are then incorporated in the food of aquatic fauna like zooplankton. The zooplankton is eaten by the higher organisms in the food chain, for example fish, which in turn are eaten by bigger fish, large birds, animals, or humans (Lynch *et al.*, 2001), with the harmful chemical concentration being amplified as it goes up the food chain. This is also known as the magnification effect.

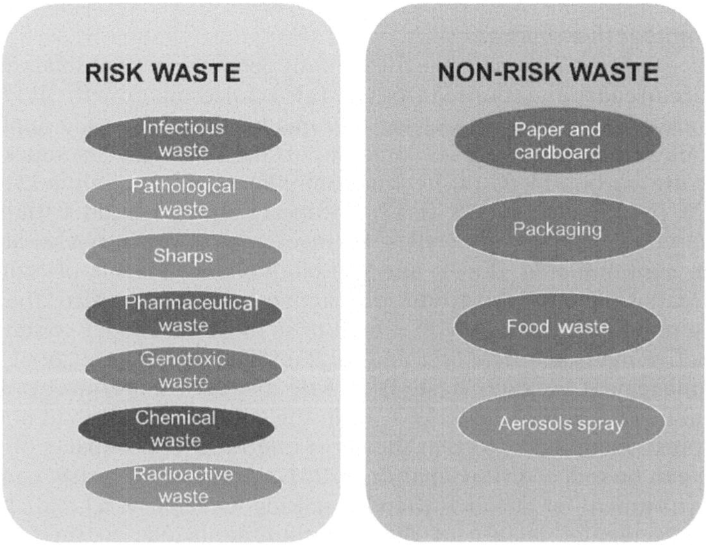

Figure 1.2 Types of hospital waste.

1.2.4 Waste management

Co-treatment of hospital as well as urban wastewaters at municipal WWTPs is not an ideal solution, since it does not provide separation of pollutants from the liquid phase, instead causing dilution of discharge into the environment (Yusuf & Athar, 2015). Pollutant concentrations are in the range of milligrams per litre, for example, 103–106 mg/L, as compared to the conventional macro-pollutants (chemical oxygen demand (COD, i.e., 75–100 mg/L), five-day biochemical oxygen demand (BOD5, i.e., 100 mg/L), phosphorus (i.e., 6 mg/L) and nitrogen (i.e., 30 mg/L) compounds) and are difficult to remove. In spite of being only at mg/L concentrations, these pollutants include a broad spectrum of compounds with significant differences in their major properties such as volatility, solubility, absorbability, polarity, biodegradability, and stability, that affect their behaviour and outcome after treatment in the WWTP. Also, the municipal WWTPs have generally been upgraded since their initial construction. The principal aim is to remove carbon, nitrogen, and phosphorus compounds, along with microbiological organisms, pollutants which arrive regularly at the WWTP in the order of mg/L and at least 106 MPN/100 mL. Conventional treatment techniques have significantly lower ability to remove micro-contaminants compared to the macro-pollutants listed (Pagliaro *et al.*, 2021).

1.3 CONCLUSIONS

Along with WHO, some of the concerned agencies have set some rules and regulations for managing hospital effluents and wastewaters. However, appropriate methods for separating hospital wastewaters at their source remain an enigma. Thus, the possibility of direct or indirect impacts of discharges to the environment cannot be ruled out. So, the crucial aspects concerning human and environmental health must be addressed urgently by the local and central authorities, and hospital management. First of all, regulation formation and implementation are essential, and appropriate measures should be taken against any bodies not following the standards. Hospital sewage and sludge with high concentrations of helminths and other pathogens requires proper treatment employing aerobic and anaerobic digestion/composting before being disposed of. Sludge that is dewatered should be trapped in the top layers of the grass roots and reeds to be successfully converted into minerals via composting. Toilet and faecal discharges of radioactive therapy patients must be separated to avoid danger from their radioactive emission. One affordable option for the collection of urine at source is Nomex technology which can enable collected urine to be transported to the pre-treatment unit where further processing can be carried out. Membrane biofilms and the activated sludge process are also highly suitable systems employed for hospital wastewater treatment, as they can remove bacteria and antibiotics to a significant extent. However, the installation, operation, and maintenance costs of these systems are comparatively higher as compared to conventional wastewater treatment. So, the technical and economic debate regarding the topic of hospital treatment plants requires further discussion and research in order to develop cost-effective and efficient solutions. Therefore, our investigations should concentrate on creating a risk-based approach to identify the optimum technologies for the treatment of hospital effluents and wastewaters. Thus, more intensive efforts should be made to revise the guidelines and plans for hospital WWT, creating regional frameworks which take into account specific local scenarios. Furthermore, the implementation and planning of these regional frameworks must be employed by all stakeholders, such as health professionals, policymakers, farmers, the general public, and so on.

ACKNOWLEDGEMENTS

The authors extend their appreciation to Universiti Teknologi PETRONAS, Malaysia and Jamia Millia Islamia, India, for providing resources in carrying out this research.

REFERENCES

Alam M. A., Ya H. H., Yusuf M., Sivraj R., Mamat O. B., Sapuan M. S., Masood F., Parveez B. and Sattar M. (2021). Modeling, optimization and performance evaluation of response surface methodology. *Materials*, **14**(16), 4703. https://doi.org/10.3390/ma14164703

Carmona-Sánchez R., Uscanga L., Bezaury-Rivas P., Robles-Díaz G., Suazo-Barahona J. and Vargas-Voräcková F. (2000). Potential harmful effect of iodinated intravenous contrast medium on the clinical course of mild acute pancreatitis. *Archives of Surgery*, **135**(11), 1280–1284. https://doi.org/10.1001/archsurg.135.11.1280

Gadupudi C. K., Rice L., Xiao L. and Kantamaneni K. (2021). Endocrine disrupting compounds removal methods from wastewater in the United Kingdom: A review. *Science*, **3**(1), 11. https://doi.org/10.3390/sci3010011

Hamid K., Shafiee P., Darbandi A., Yusuf M., Mahmoudi S., Goudarzi Z. M. and Mirzamohammadi S. (2021). Application of polyoxometalate-based composites for sensor systems: A review. *Journal of Composites and Compounds*, **3**(7), 129–139. https://doi.org/10.52547/jcc.3.2.6

Hernando M. D., Agüera A. and Fernández-Alba A. R. (2007). LC-MS analysis and environmental risk of lipid regulators. *Analytical and Bioanalytical Chemistry*, **387**(4), 1269–1285. https://doi.org/10.1007/s00216-006-0781-y

Iqbal F., Abdullah B., Oladipo H., Yusuf M., Alenazey F., Nguyen T. D. and Ayoub M. (2021). Recent developments in photocatalytic irradiation from CO2 to methanol. In: Nanostructured Photocatalysts- From Fundamental to Practical Applications. Elsevier, pp. 519–540. https://doi.org/10.1016/B978-0-12-823007-7.00015-8

Joss A., Zabczynski S., Göbel A., Hoffmann B., Löffler D., McArdell C. S., Ternes T. A., Thomsen A. and Siegrist H. (2006). Biological degradation of pharmaceuticals in municipal wastewater treatment: proposing a classification scheme. *Water Research*, **40**(8), 1686–1696. https://doi.org/10.1016/j.watres.2006.02.014

Jureczko M. and Kalka J. (2020). Cytostatic pharmaceuticals as water contaminants. *European Journal of Pharmacology*, **866**, 172816. https://doi.org/10.1016/j.ejphar.2019.172816

Khan A. H., Aziz H. A., Khan N. A., Hasan M. A., Ahmed S., Farooqi I. H., Dhingra A., Vambol V., Changani F., Yousefi M., Islam S., Mozaffari N. and Mahtab M. S. (2021a). Impact, disease outbreak and the eco-hazards associated with pharmaceutical residues: a critical review. *International Journal of Environmental Science and Technology*, **19**, 677–688, https://doi.org/10.1007/s13762-021-03158-9

Khan N. A., Ahmed S., Vambol V. and Vambol S. (ed.) (2021b). Pharmaceutical Wastewater Treatment Technologies: Concepts and Implementation Strategies. IWA Publishing, London, UK. https://doi.org/10.2166/9781789061338

Kozub P., Kozub S., Mozaffari N., Sihag P., Mozaffari N., Vambol S., Vambol V., Ramsawak N. and Khan N. A. (2021). Treatment schemes – conventional and dedicated for PhACs treatment. In: Pharmaceutical Wastewater Treatment Technologies: Concepts and Implementation Strategies. IWA Publishing. https://doi.org/10.2166/9781789061338_0181

Kumari A., Maurya N. S. and Tiwari B. (2020). Hospital wastewater treatment scenario around the globe. In: Current Developments in Biotechnology and Bioengineering. p. 7728. https://doi.org/10.1016/B978-0-12-819722-6.00015-8; https://www.sciencedirect.com/book/9780128197226/current-developments-in-biotechnology-and-bioengineering

Laffite A., Kilunga P. I., Kayembe J. M., Devarajan N., Mulaji C. K., Giuliani G., Slaveykova V. I. and Poté J. (2016). Hospital effluents are one of several sources of metal, antibiotic resistance genes, and bacterial markers disseminated in sub-saharan urban rivers. *Frontiers in Microbiology*, **7**, 1–14. https://doi.org/10.3389/fmicb.2016.01128

Lynch J. M., Wiseman A. and De Leij F. A. A. M. (2001). Ecotoxicology. In: Encyclopedia of Biodiversity, 2nd edn, Vol. **2**, pp. 118–126, Elsevier, UK. https://doi.org/10.1016/B978-0-12-384719-5.00043-5

Mousavi S. A. and Khodadoost F. (2019). Effects of detergents on natural ecosystems and wastewater treatment processes: a review. *Environmental Science and Pollution Research*, **26**(26), 26439–26448. https://doi.org/10.1007/s11356-019-05802-x

Ngigi A. N., Magu M. M. and Muendo B. M. (2020). Occurrence of antibiotics residues in hospital wastewater, wastewater treatment plant, and in surface water in Nairobi county, Kenya. *Environmental Monitoring and Assessment*, **192**(1), 1–16. https://doi.org/10.1007/s10661-019-7952-8

Pagliaro M., Ciriminna R., Yusuf M., Eskandarinezhad S., Wani I. A., Ghahremani M. and Nezhad Z. R. (2021). Application of nanocellulose composites in the environmental engineering as a catalyst, flocculants, and energy storages: A review. *Journal of Composites and Compounds*, **3**(7), 114–128. https://doi.org/10.52547/jcc.3.2.5

Ravichandran R. (2017). Management of radioactive wastes in a hospital environment. In: Modelling Trends in Solid and Hazardous Waste Management. pp. 1–14, Springer Singapore, Singapore. https://doi.org/10.1007/978-981-10-2410-8

Rosdin R. D. B., Yusuf M. and Abdullah B. (2021). Dry reforming of methane over Ni-based catalysts: effect of ZrO_2 and MgO addition as support. *Materials Letters: X*, **12**, 100095. https://doi.org/10.1016/j.mlblux.2021.100095

Schröder P., Helmreich B., Škrbić. B., Carballa M., Papa M., Pastore C., Emre Z., Oehmen A., Langenhoff A., Molinos M., Dvarioniene J., Huber C., Tsagarakis K. P., Martinez-Lopez E., Pagano S., Meric Vogelsang C. and Mascolo G. (2016). Status of hormones and painkillers in wastewater effluents across several European states – considerations for the EU watch list concerning estradiols and diclofenac. *Environmental Science and Pollution Research*, **23**(13), 12835–12866. https://doi.org/10.1007/s11356-016-6503-x

Seiler C. and Berendonk T. U. (2012). Heavy metal driven co-selection of antibiotic resistance in soil and water bodies impacted by agriculture and aquaculture. *Frontiers in Microbiology*, **3**, 1–10. https://doi.org/10.3389/fmicb.2012.00399

US Environmental Protection Agency (2015). National pollutant discharge elimination system (NPDES) electronic reporting rule. *Federal Register*, **80**(204), 1–96. Available at: www.epa.gov/dockets

Vambol S., Vambol V., Mozaffari N., Mahmoud A. E. D., Ramsawak N., Mozaffari N., Ziarati P. and Khan N. A. (2021a). Comprehensive insights into sources of pharmaceutical wastewater in the biotic systems. In: Pharmaceutical Wastewater Treatment Technologies: Concepts and Implementation Strategies. IWA Publishing. https://doi.org/10.2166/9781789061338_0017

Vambol V. (2021b). Legislative and criminal law aspects of water protection and prevention of pharmaceuticals accumulation in nature. In: Pharmaceutical Wastewater Treatment Technologies: Concepts and Implementation Strategies. IWA Publishing. https://doi.org/10.2166/9781789061338_0113

Yusuf M. (2016). Natural Gas hydrates: The future's fuel. *Journal of Environmental Research And Development*, **10**(4), 738–746.

Yusuf M. and Athar M. (2015). Biodiesel production using hexane as Co-solvent. *Journal of Biofuels*, **6**(2), 88. https://doi.org/10.5958/0976-4763.2015.00013.6

Yusuf M., Farooqi A. S., Alam M. A., Keong L. K., Hellgardt K. and Abdullah B. (2020). Latest trends in syngas production employing compound catalysts for methane dry reforming. *IOP Conference Series: Materials Science and Engineering*, **991**(1), 12071. https://doi.org/10.1088/1757-899X/991/1/012071

Yusuf M., Farooqi A. S., Alam M. A., Keong L. K., Hellgardt K. and Abdullah B. (2021a). Performance of Ni/Al_2O_3-MgO catalyst for Dry reforming of methane: effect of preparation routes. *IOP Conference Series: Materials Science and Engineering*, **1092**(012069), 1–9. https://doi.org/10.1088/1757-899X/1092/1/012069

Yusuf M., Beg M., Ubaidullah M., Shaikh S., Keong L., Hellgardt K. and Abdullah B. (2021b). Kinetic studies for DRM over high-performance Ni-W/Al_2O_3-MgO catalyst. *International Journal of Hydrogen Energy*, 1–10. https://doi.org/10.1016/j.ijhydene.2021.08.021

Yusuf M., Salam A., Alam M. A., Keong L. K., Hellgardt K. and Abdullah B. (2021c). Response surface optimization of syngas production from greenhouse gases via DRM over high performance Ni-W catalyst. *International Journal of Hydrogen Energy*, 1–14. https://doi.org/10.1016/j.ijhydene.2021.05.153

Yusuf M., Salam A., Al-kahtani A., Ubaidullah M., Alam M. A., Keong L. K., Hellgardt K. and Abdullah B. (2021d). Syngas production from greenhouse gases using Ni–W bimetallic catalyst via dry methane reforming: effect of W addition. *International Journal of Hydrogen Energy*, **46**(53), 27044–27061. https://doi.org/10.1016/j.ijhydene.2021.05.186

Yusuf M., Alnarabiji M. S. and Abdullah B. (2021e). Clean hydrogen production technologies BT. In: Advances in Sustainable Energy: Policy, Materials and Devices, Y. Gao (ed.), Springer International Publishing, Cham, Switzerland, pp. 159–170. https://doi.org/10.1007/978-3-030-74406-9_5

Zotesso J. P., Cossich E. S., Janeiro V. and Tavares C. R. G. (2017). Treatment of hospital laundry wastewater by UV/H_2O_2 process. *Environmental Science and Pollution Research*, **24**(7), 6278–6287. https://doi.org/10.1007/s11356-016-6860-5

doi: 10.2166/9781789062625_0013

Chapter 2

Pharmaceuticals measurements and estimation methods

Nastaran Mozaffari[1], Ştefan Ţălu[2], Niloofar Mozaffari[3], Mika Sillanpää[4,5,6], Viola Vambol[7] and Sergij Vambol[8]*

[1]Department of Chemical Engineering, Universite Laval, Quebec, QC G1V 0A6, Canada
[2]The Directorate of Research, Development and Innovation Management (DMCDI), Constantin Daicoviciu Street, no. 15, Cluj-Napoca, 400020, Cluj County, Romania
[3]Department of Chemical Engineering, Universite Laval, Quebec, QC G1V 0A6, Canada
[4]Department of Chemical Engineering, School of Mining, Metallurgy and Chemical Engineering, University of Johannesburg, P. O. Box 17011, Doornfontein 2028, South Africa
[5]Faculty of Science and Technology, School of Applied Physics, University Kebangsaan Malaysia, 43600, Bangi, Selangor, Malaysia
[6]International Research Centre of Nanotechnology for Himalayan Sustainability (IRCNHS), Shoolini University, Solan, 173212, Himachal Pradesh, India
[7]Department of Applied Ecology and Environmental Sciences, National University "Yuri Kondratyuk Poltava Polytechnic", Poltava, Ukraine
[8]Department of Life Safety, State Biotechnological University, Kharkiv, Ukraine
*Corresponding author. Niloofar.mozaffari1991@gmail.com

ABSTRACT

Health centers, particularly hospitals, produce 200–1200 L/bed volume of wastewater derived from hospitalized patients/day, in which detrimental contaminants such as remains of pharmaceuticals, chemical materials, microbial pathogens, radioisotopes, and viral matters like SARS-CoV-2 are existent. It is a well-known fact that the accumulation of both hospital and urban wastewater together can raise the concentration of harmful pollutants. Thus, it is essential to provide a technology to detect pharmaceuticals in water and wastewater. Biosensors, including carbon nanotubes, graphene-based sensors, gold nanoparticles-based sensors, and biosensors based on molecularly imprinted polymers are well-known in pharmaceutical detection due to their unique properties. In this chapter, the function of various biosensors in the detection of pharmaceutical water pollution is explored. Carbon nanotube sensors possess high sensitivity for pharmaceutical detection. Graphene was reported as the best sensor when compared with carbon nanotubes because of its low price and large surface area. It can detect a wide range of pharmaceuticals, due to their extensive ranges of chemical modification and bimolecular binding schemes. Carbon black is another carbon-based sensor used to detect pharmaceuticals owing to its low price, large surface area, and high stability. Among nanoparticles, gold nanoparticles are known as biosensors for microbial and pharmaceutical contaminants detection in aquatic environments.

Editor(s): Nadeem A Khan, Viola Vambol, Sergij Vambol, Nastaran Mozaffari and Niloofar Mozaffari

Keywords: Harmful pollutants, Hospitals, Technology to detect pharmaceuticals, Wastewater

CONTENTS

2.1 INTRODUCTION

Health centers, particularly hospitals, can produce 200–1200 L/bed volume of wastewater derived from hospitalized patients/day. Actually, various factors including medical proficiency, bed capacity, weather condition and climate, geographical situation, quality of provided services by health centers (e.g., laundry, bathroom, and kitchen) can contribute to increases and decreases in the wastewater volume charged in hospitals (Verlicchi *et al.*, 2010a, 2010b). Contents of hospital wastewater usually include a broad span of detrimental contaminants such as remains of pharmaceuticals, chemical materials, microbial pathogens, radioisotopes, and viral matters like SARS-CoV-2 (Achak *et al.*, 2020; Basturka *et al.*, 2020; El-Ogri *et al.*, 2016; Verlicchi *et al.*, 2010a, 2010b). There are two principal categories of contaminated matters in hospital wastewater, which are shown in Table 2.1.

Despite the fact that many chemical, biological, and physical threats to public and environmental health can be caused by these substances, in many countries they are often discharged into receiving aquatic environments without any purification (Leprat, 1998). In addition, the quantity of pharmaceuticals in waste streams should be counted since the Food and Drug Administration

Table 2.1 Two main categories of hospital wastewater content (Achak *et al.*, 2020)

Polluted Content of Hospital Wastewater	
Macro-pollutants	**Micro-pollutants**
Physicochemical factors such as chemical oxygen demand (COD), pH, total organic carbon (TOC), biological oxygen demand (BOD), suspended solids (SS), chloride, ammonium ions.	Absorbable organic halogens (AOX) Cytostatics Analgesics Hormones Cleansers
Microbiological factors such as bacteria (Lococcus, Enterococcus, Salmonella, Shigella), coliforms, and viruses.	Disinfectants Phenols Contrast materials Heavy metals including cadmium, chrome, nickel, copper, zinc, iron, and lead Antibiotics including ciprofloxacin, and sulfamethoxazole

(FDA) has not yet estimated the exact quantity of drugs and their compounds in wastewater, particularly wastewater originating from hospitals (National Center for Biotechnology Information, 2012). Due to the fact that aquatic environments have direct and consistent contact with contaminated water, marine ecosystems and habitats face a higher risk than human beings from contaminated waste streams. It is significant to note that different types of organisms show various reactions to pharmaceutical elements depending on how their mechanisms process them (Environmental Protection Agency, 2010). For instance, in published literature, it was proved that male fishes created a special protein that is fabricated for egg production only by female fishes (Gilbert, 2012). Due to the absence of permanent and extended exposure to the polluted water, the levels of human exposure to these pharmaceuticals may vary (Environmental Protection Agency, 2010). According to the results of drinking water testing in some regions, parts per million or trillion levels of pharmaceuticals such as ibuprofen, sulfamethoxazole, and carbamazepine have been found in drinking water, and, also, reusing of treated wastewater is encouraged due to water scarcity issues (Deziel, 2014; Luo *et al.*, 2014).

A large number of recent studies revealed that the fecal-oral route is another means of SARS-CoV-2 transmission, although SARS-CoV-2 is able to transfer between individuals through respiratory droplets via close contact as well (Ling *et al.*, 2019; Wu *et al.*, 2020; Yang *et al.*, 2020; Zhang *et al.*, 2020a, 2020b, 2020c). According to studies, viruses can survive and become more resistant in sewage if the fecal matter of SARS-infected patients is discharged to it (Casanova *et al.*, 2020; Hung, 2003; Leung *et al.*, 2003). Regarding previous works in the literature, the existence of SARS-CoV-2 in wastewater has been found by many scientists in various geographic regions, and its survival can depend on temperature, retention period, pH, organic matter, and so on. (Wang *et al.*, 2005). If SARS-CoV-2 transfers through the fecal route into the aquatic environment, especially from health center wastewater, then effective treatment strategies and techniques should be provided, particularly in countries with inadequate sanitation and diagnostic systems, such as Africa, due to the fact that these viral matters can survive and have a long-term life span in sewage derived from hospitals (Lodder & de Roda Husman, 2020).

Two source types, namely point- and non-point- sources release pharmaceuticals and personal care products (PPCPs) into aquatic ecosystems. There are some similarities between hospital and urban wastewaters in many countries, based on types of contaminants, volume, and concentrations, especially if they are released into urban sewage, and finally gathered to a wastewater purification facility (Figure 2.1) (Leprat, 1998). In addition, previous studies revealed that charged contaminant in hospital effluent is higher than in wastewater derived from urban sources, in terms of the evaluation of physical, chemical, and microbiological characteristics of both wastewater sources. According to comparative review studies, the accumulation of both hospital and urban wastewater together can raise the concentration of pollutants, since hospital wastewater contains COD, BOD, and SS, at levels 2–3 times higher than in municipal wastewater (Mesdaghinia *et al.*, 2009; Tsakona *et al.*, 2007; Verlicchi *et al.*, 2010a, 2010b, 2012) (Table 2.2). Thus, it is essential to provide a

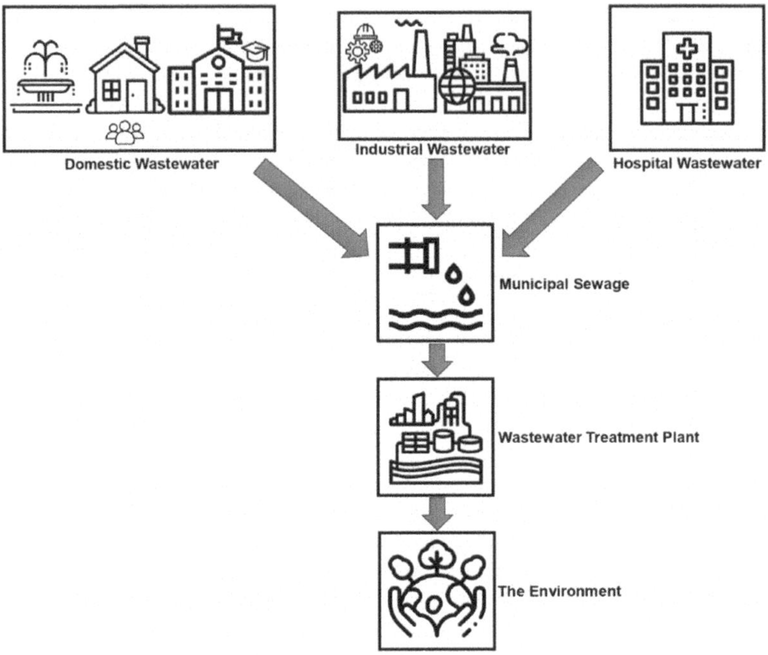

Figure 2.1 A treatment cycle of three different types of wastewaters: domestic/industrial/hospital wastewater (Leprat, 1998).

Table 2.2 Main characteristics of hospital/urban wastewater (Mesdaghinia *et al.*, 2009; Tsakona *et al.*, 2007; Verlicchi *et al.*, 2010a, 2010b, 2012)

Principal Properties of Hospital/Urban Wastewater				
Hospital Wastewater				
Physicochemical Parameters				
pH	COD (mg L^{-1})	BOD_5 (mg L^{-1})	SS (mg L^{-1})	Chlorides (mg L^{-1})
6.9–9.18	450–2,300	150–603	120–400	80–400
Microbiological Parameters				
Total coliforms (MPN/100 mL)		Fecal coliforms (MPN/100 mL)		*E. coli* (MPN/100 mL)
10^6–10^9		10^3–10^7		10^3–10^6
Urban Wastewater				
Physicochemical Parameters				
pH	COD (mg L^{-1})	BOD_5 (mg L^{-1})	SS (mg L^{-1})	Chlorides (mg L^{-1})
7.5–8.5	500–600	100–400	120–350	30–100
Microbiological Parameters				
Total coliforms (MPN/100 mL)		Fecal coliforms (MPN/100 mL)		*E. coli* (MPN/100 mL)
10^7–10^8		10^6–10^7		10^6–10^7

committed pre-treatment section for hospital wastewater before discharging it (Verlicchi *et al.*, 2010a, 2010b).

Research on pharmaceuticals, illegal drugs, industrial chemical matters, biological elements, and nutrients is receiving significant attention in various fields of study. Also, the determination of chlorinated paraffins (CPs) in wastewater has been investigated by several review articles (Kümmerer, 2009; Lis *et al.*, 2021; Maier *et al.*, 2001; Nikolaou *et al.*, 2007; Zhang *et al.*, 2010). Generally, there are many different technologies which can be used to calculate the quantity of pharmaceuticals in water as well as wastewater. One of these is biosensors. Various kinds of biosensors are available, including those based on carbon nanotubes (CNTs), and molecularly imprinted polymers (MIPs), graphene-based sensors, gold nanoparticles (AuNPs) based sensors, and so on., which each have their unique structure, performance, benefits, and disadvantages, that will be explained comprehensively in this chapter.

2.2 TECHNOLOGIES – BIOSENSORS

Biosensors as analytical systems, combine biological or biomimetic materials with a physical-chemical or a microsystem transducer (González-Martínez *et al.*, 1999). Although several subsidiary stages are required for conventional sensors to add reagents, bioassays automate their process in order to be further developed into novel biosensors (Patel, 2002; Rogers, 2006). Biosensors have advantages over conventional techniques or batch systems, such as being cheaper, having a high-level capacity of detection, and using robotic analytic procedures (Onnerfjord *et al.*, 1998). One of the main superiorities of biosensors is their ability to detect both pharmaceuticals and microbial pollutants with the use of lower numbers of samples and reagents, in addition to having a high capacity for pollutants detection even if their concentration is very low (Samendra *et al.*, 2014). Table 2.3 demonstrates two major classifications of biosensors based on certain elements for the identification of the analytes, and their signal transducer. An overview of various biosensors developed for environmental applications, and their performance principles are presented in Table 2.4.

One of the well-known biosensors is based on carbon nanotubes (CNTs), classified according to its ability to align with various molecules readily, and possesses well-identified and evaluated electrical transmission characteristics

Table 2.3 Two main classifications of biosensors depending on identifiers of the analytes, and their signal converter (Adegoke *et al.*, 2019)

Given Factors for Recognizing the Analytes	Signal Converter
Immuno-chemical biosensors Enzymatic biosensors Entire-cell biosensors Non-enzymatic receiver biosensors DNA biosensors	Electro-chemical, thermal, spectroscopic, piezoelectric, and surface-wave technologies are techniques that the physical transducer is based on.

Table 2.4 Diverse biosensors used for environmental applications and their properties

Type of Biosensor	Biosensors for withholding of acetylcholinesterase	Probe-based/ genetically-encoded/ fluorescence-tagged biosensors (e.g., GFP-based bacterial biosensor).*	Nanomaterials-based biosensors	Microbial fuel cell-based biosensor
Origin	Electrochemistry	Fluorescence	Including a great range of types, such as electrochemical, visual, fluorescence, optical, etc.	Optical
Particular Usage	Contamination and impact of pesticide	Bioaccumulation	Detecting various pollutants through a proper aptamer, including pharmaceuticals, microbes, etc.	Monitoring eco-toxicity through heavy metals, PPCPs, other related pollutants like BOD.
Reference	Pundir and Chauhan (2012)	Priyadarshi *et al.* (2012)	Da Costa Silva *et al.* (2013), Govindhan *et al.* (2014), Guo (2013), Hutter and Maysinger (2013), Sang *et al.* (2015), Senveli and Tigli (2013), Valentini *et al.* (2013)	Sun *et al.* (2015)

**Note:* The green fluorescent protein (GFP) is a protein originated from jellyfish that exhibits fluorescence once it is exposed to light.

(Ajayan, 1999). The influence of molecular modifications in CNTs can be used to reflect the electrical characteristics, since most atoms in CNTs seem to be surface-based (Yang *et al.*, 2007). One of the advantages of CNTs is their high level of sensitivity due to their large surface area to volume ratio associated with an array of binding places for a large series of nanoparticles and molecules (Ali *et al.*, 2007; An *et al.*, 2004). Activity as well as the large surface area to volume ratio, and a large sorption efficiency are unique properties of CNTs that make them suitable for use in wastewater treatment, and also, they are commonly used as a biosensor for microbial pollutants detection, particularly antibiotic-resistant microbial matters, in drinking water sources (Ong *et al.*, 2010). The surface acidity of CNTs contributes to their performance as a good adsorbent as well as a favorable detector or sensor (Rao *et al.*, 2007). Some researchers investigated the ability of CNTs to act as a biosensor, through optimizing two potential single-walled CNTs structures as immunosensors, for sensing the bacteriophage MS_2 (a virus existing in sewage). The ease of application of biosensors is acknowledged by the positive results of these studies with the two detection methods used (Davis *et al.*, 2010; Rao *et al.*, 2007). Currently, biosensors can be applied for the identification of both viruses and bacteria in the aquatic environment in less than 1200 s, and this is possible through the advent of Fourier transform infrared spectroscopy (FTIR) technology which can detect *E. coli* in just 6 hours (Davis *et al.*, 2010) (Table 2.5). Table 2.6 presents characteristics of some biosensors used for pharmaceuticals detection.

Coating various nanomaterials onto biosensors can enable them to detect pharmaceuticals and bacterial contaminants in water. Graphene as a monolayer 2D allotrope of carbon atoms (honeycomb-like 2D arrangement), is one of the nanomaterials which can improve the ability of biosensors because of its distinctive nano frameworks and other remarkable characteristics. It is a reference point for novel applications because of its significant usage in nanocomposites, nanoelectronics, optoelectronic facilities, electrochemical supercapacitor systems, and manufactured field-effect transistors. Unique characteristics including special optical characteristics, high electrical conductivity, large surface to volume ratio, noticeable carrier density, great carrier mobility, and considerable mechanical strength have made graphene a popular substance for use in optoelectronic systems, supercapacitors, and other sensors (Adegoke *et al.*, 2019). Graphene's low cost, large surface area to volume ratio, and homogeneous structure have made graphene more suitable than carbon nanotubes for biosensors. In addition, graphene-based sensors (GBS) are able to detect a wide range of pharmaceuticals, due to their extensive range of chemical modification and bimolecular binding schemes (Soldano *et al.*, 2010). The antifouling ability of graphene-based electrodes is helpful for wastewater analysis, and hence, GBS can be applied for monitoring and detecting pharmaceuticals, or PPCPs, over-the-counter (OTC) drugs, cholesterol, glutamate, organic matters, and protein biomarkers, properly (Adegoke *et al.*, 2019). Chen *et al.* (2015) developed a quick and exact electroanalytical technique to evaluate the existence of ciprofloxacin in

Table 2.5 Microbial pollutants detection in water through some applications of biosensors

Microbial Matter	*Escherichia coli* (tolC strains)	*E. coli* (tolC strains)	*E. coli* (tolC strains)	*Pseudomonas* sp. *Moraxella* sp.	Recombinant luminescent bacteria	*Spirulina subsalsa*
Application	Mitomycin C	Formaldehyde	Cerulinin	p-Nitrophenol	Organotin compounds	Chlorophenols
Reference	Norman *et al.* (2005)	Norman *et al.* (2005)	Bechor *et al.* (2002)	Zhang *et al.* (2012)	Durand *et al.* (2003)	Tonnina *et al.* (2002)

Table 2.6 Characteristics of several biosensors used for pharmaceuticals identification in the aquatic environment

Analyte	Biosensor	Method of Detection	Linear Range (μmol L^{-1})	Sensitivity	Sample	Contaminant	Ref
Acetaminophen	Au/MWCNT-CuNPs	DPV	0.5–80	2.79	Wastewater river water	Anti-Inflammatories	Daneshvar *et al.* (2016)
Diclofenac	GCE/Cu(OH)2 -IL-MWCNT		0.18–119	0.015	Seawater fish serum		Arvand *et al.* (2012)
Ciprofloxacin	BDD/MWCNT-nafion		0.005–0.05/0.05–10	2.08	Wastewater	Antibiotics	Gayen and Chaplin (2016)
Sulfamethoxazole	CPE (MWCNT-SbNPs)		0.1–0.7	0.57	Natural water (dam)		Cesarino *et al.* (2013)

MWCNT: multi-walled carbon nano tube; DPV: different pulse voltammetry; GCE: glassy carbon electrode; BDD: boron-doped diamond; CPE: Carbon paste electrodes.

see creatures' bodies. The process was implemented using a graphene oxide (GO)-based sensor diffused in a hybrid matrix of sodium polyacrylate-chloropalladic acid, with different pulse voltammetry (DPV) applied. Compared to other electrochemical sensors based on multi-walled carbon nanotubes (MWCNT), $MgFe_2O_4$-based MWCNT, β-cyclodextrin/L-arginine carbon pasted electrode, horseradish peroxidase, poly (alizarin red), the results of this research study show a broad linear range and high sensitivity. In addition, this research team (Chen *et al.*, 2015) applied an analytical tool in the form of reduced graphene oxide (rGO)-modified screen-printed electrodes (SPE), in order to provide a promising substitute for point-of-care monitoring of sulfamethoxazole. Since ascorbic acid reagent can reduce the stacking and van der Waals interactions between graphene flakes, this team applied ascorbic acid as a GO reducing agent. Table 2.7 represents several graphene-based biosensors used for pharmaceuticals detection in water, and their characteristics (Torrinha *et al.*, 2020).

Another carbon-based nanomaterial is carbon black (CB) which is known for use in manufacturing various types of sensors due to its unique properties, including being economical, having an enormous surface area, high conductivity, and high stability, and being able to create films with homogeneous structures. Carbonaceous materials that have a bulkier framework are manufactured through the combustion of hydrocarbons. Their compositions contain aggregation or agglomeration of spherical particles in which every particle possesses a turbostratic structure of random packing layers of graphite. Delgado *et al.* (2018) manufactured a sensor made of CB nanoballs and potato starch deposited onto a glassy carbon electrode (GCE), for detection of tetracycline in tap water and river water. The results indicated a suitable sensitivity of this sensor due to its increased conductivity and large surface area. Generally, carbon nanomaterials have low conductivity in comparison to metals, but they are associated with a metallic or semiconductive (10–4 Ω cm) behavior in order to achieve high transfer rates of the electron. The production and commercial availability of carbon nanomaterials at a reasonable price may be readily achievable, through the relatively simple processing of carbon, an almost plentiful chemical element. The complete process of pharmaceutical detection in water through the application of nanomaterials is presented in Figure 2.2. Figure 2.3 shows the principal advantages of carbon-based nanomaterials leading to their significant-range application in the technology of (bio)sensors (Torrinha *et al.*, 2020).

Molecularly imprinted polymers (MIPs) are another type of biosensor. Bui and Haupt (2010) described MIPs as artificial substances with particular cavities designed for a target molecule, and the target analyte was identified by these materials, through their affinities and selectivity which is comparable to those of enzyme–substrate, ligand–receptor, and antibody–antigen interactions. The interaction between the template molecule and complementary functional monomer group can determine the effectiveness of the MIP sensors (Whitcombe *et al.*, 2011). Despite the fact that detection limits can sometimes be less than half of the permitted concentration for elements in water, the minimum sample required for some MIP biosensors is often 2.5 mL with an experiment time of two minutes. Compatibility of the

Table 2.7 Properties of several graphene-based biosensors applied for pharmaceutical identification in the aquatic environment

Analyte	Biosensor	Method of Detection	Linear Range (μmol L^{-1})	Sensitivity	Sample	Contaminant	Ref
Ciprofloxacin	GCE/Pd-PAAS-GO	DPV	0.18–10.8	0.867	Shrimp	Antibiotics	Chen *et al.* (2015)
Tetracycline	GCE/graphene/L-cysteine		8.0–140	0.027	River water		Sun *et al.* (2017)
Sulfamethazine	GCE/rGO-AuNPs		0.5–6.5	0.32	Wastewater		Silva and Cesarino (2019)
Piroxicam	GCE/ZnO-GO-glutathione	DPV	0.1–100/100–500	0.206/0.030	Tap water	Anti-inflammatories	Dhanalakshmi *et al.* (2020)
Diclofenac	GCE/graphene-PDDA		20–100/20–200	0.071/0.636	Lake water		Okoth *et al.* (2018)
17β-Estradiol	GCE/rGO-MIP-Fe3O4	DPV	0.05–10	0.871	Water	Hormones	Li *et al.* (2015)
Diethylstilbestrol	GCE/exfoliated graphene		0.010–1.5	4.43	Lake water		Hu *et al.* (2015)

PAAS: sodium polyacrylate; PDDA: poly (diallyldimethylammonium chloride).

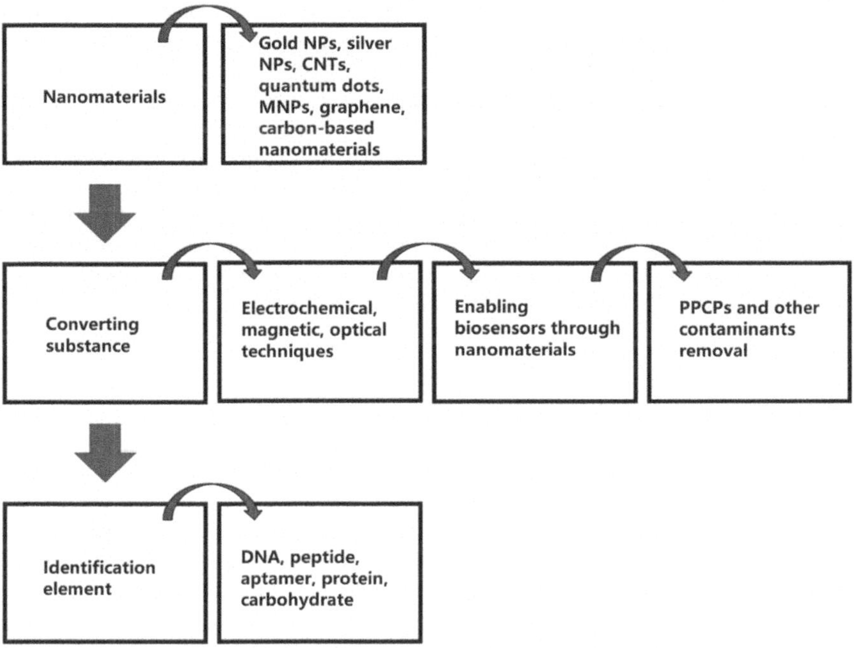

Figure 2.2 The general process of polluted elements identification in water through nanomaterials coated on biosensors. MNPs: magnetic nanoparticles (Torrinha *et al.*, 2020).

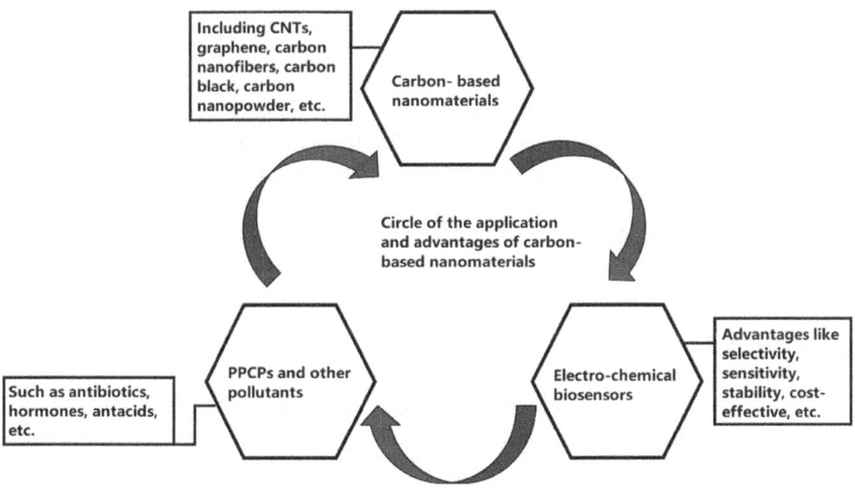

Figure 2.3 Application and benefits of carbon-based nanomaterials for pharmaceutical detection in water (Torrinha *et al.*, 2020).

equipment with water is one of the challenges of MIP biosensors, thus post-modification of the MIPs to make their surface hydrophilic is essential. In this regard, factors such as size, control of the binding sites, and the entire processes of the system have been improved appreciably. Gold nanoparticles (AuNPs) are another type of NP that can improve the ability of biosensors for microbial and pharmaceutical pollutants detection in aquatic environments. The attachment of the nanoparticles to the biological or PPCPs' molecules can be conducted effectively through modification of their surfaces. AuNPs and gold nanorods as magnetic nanoparticles (MNPs) are identified as a potential candidate for the timely and efficient detection of toxins, heavy metals, and inorganic and organic matters from pharmaceutical sources (Wang and Yu, 2013).

2.3 CONCLUSION

A range of 200–1200 L/bed volume of wastewater is being produced in health centers every day. Despite the fact that many chemical, biological, and physical threats to public and environmental health can be caused by these substances, they are released to the aquatic environments in urban areas or directly onto surface water without any purification in many countries. The accumulation of both hospital and domestic wastewater together can raise the concentration of pollutants. Thus, it is essential to provide a committed pre-treatment section for hospital wastewater before discharging it. Generally, there are many different technologies, one of which is biosensors, which can be used to detect the quantity of pharmaceuticals in water and wastewater. One of the important superiorities of biosensors is their ability to detect both pharmaceuticals and microbial pollutants with the use of lower numbers of samples and reagents, in addition to having a high capacity for pollutants detection even if their concentration is very low. One of the advantages of carbon nanotubes is their high level of sensitivity due to their large surface area to volume ratio associated with an array of binding places for a large series of nanoparticles and molecules. Superior properties of graphene, like special optical and electrical properties, large surface area, great carrier mobility, and considerable mechanical strength have made graphene a popular sensor for the detection of pharmaceutical water pollution. Among carbon-based materials, carbon black is considered the best sensor due to its low-price, large surface area, good conductivity, and high stability. Among nanoparticles, gold nanoparticles are known as biosensors for microbial and pharmaceutical contaminants detection in aquatic environments.

REFERENCES

Achak M., Bakri S. A., Chhiti Y., Alaoui F. E. M., Barka N. and Boumya W. (2020). SARS-CoV-2 in hospital wastewater during outbreak of COVID-19: A review on detection, survival and disinfection technologies. *Science of the Total Environment*, **761**, 143192.

Adegoke A. A., Singh G. and Stenstrom T. A. (2019). Biosensors for monitoring pharmaceutical nanocontaminants and drug resistant bacteria in surface water, subsurface water and wastewater effluent for reuse. *Nanoparticles in Pharmacotherapy*, 525–559, https://doi.org/10.1016/B978-0-12-816504-1.00014-4.

Ajayan P. M. (1999). Nanotubes from carbon. *Chemical Reviews*, **99**, 1787–1800.

Ali S. R., Ma Y., Parajuli R. R., Balogun Y., Lai W. Y.-C. and He H. (2007). A nonoxidative sensor based on a self-doped polyaniline/carbon nanotube composite for sensitive and selective detection of the neurotransmitter dopamine. *Analytical Chemistry*, **79**, 2583–2587.

An K. H., Jeong S. Y., Hwang H. R. and Lee Y. H. (2004). Enhanced sensitivity of a gas sensor incorporating single-walled carbon nanotube polypyrrole nanocomposites. *Advanced Materials*, **16**, 1005–1009.

Arvand M., Gholizadeh T. M. and Zanjanchi M. A. (2012). MWCNTs/$Cu(OH)_2$ nanoparticles/IL nanocomposite modified glassy carbon electrode as a voltammetric sensor for determination of the non-steroidal anti-inflammatory drug diclofenac. *Materials Science and Engineering: C*, **32**, 1682–1689.

Basturka I., Varankb G., Murat-Hocaoglua S., Yazici-Guvenc S., Oktem-Olguna E. and Canli O. (2020). Characterization and treatment of medical laboratory wastewater by ozonation: optimization of toxicity removal by central composite design. *Ozone: Science & Engineering*, **43**(3), 213–227.

Bechor O., Smulski D. R., Van Dyk T. K., La Rossa R. A. and Belkin S. (2002). Recombinant microorganisms as environmental biosensors: pollutants detection by Escherichia coli bearing fabA: lux fusions. *Journal of Biotechnology*, **94**, 125–132.

Bui B. T. S. and Haupt K. (2010). Molecularly imprinted polymers: synthetic receptors in bioanalysis. *Analytical and Bioanalytical Chemistry*, **398**(6), 2481–2492.

Casanova L., Rutala W. A., Weber D. J. and Sobsey M. D. (2020). Survival of surrogate coronaviruses in water. *Water Research*, **43**, 1893–1898.

Cesarino I., Cesarino V. and Lanza M. R. V. (2013). Carbon nanotubes modified with antimony nanoparticles in a paraffin composite electrode: simultaneous determination of sulfamethoxazole and trimethoprim. *Sensors and Actuators B: Chemical*, **188**, 1293–1299.

Chen H., Bo L., He M., Mi G. and Li J. (2015). Synthesis and application of graphene–sodium polyacrylate–Pd nanocomposite in amperometric determination of ciprofloxacin. *Chemistry Letters*, **44**, 815–817.

Da Costa Silva L. M., dos Santos S. V. P., Salgado A. M. and Pereira K. S. (2013). Biosensors for contaminants monitoring in food and environment for human and environmental health. In: State of the Art in Biosensors – General Aspects. *In Tech*, 151–168, https://doi.org/10.5772/55617.

Daneshvar L., Rounaghi G. H. and Tarahomi S. (2016). Voltammetric paracetamol sensor using a gold electrode made from a digital versatile disc chip and modified with a hybrid material consisting of carbon nanotubes and copper nanoparticles. *Microchimica Acta*, **183**, 3001–3007.

Davis R., Irudayaraj J., Reuhs B. L. and Mauer L. J. (2010). Detection of E. coli O157:H7 from ground beef using Fourier transform infrared (FT-IR) spectroscopy and chemometrics. *Journal of Food Science*, **75**, M340–M346.

Delgado K. P., Raymundo-Pereira P. A., Campos A. M., Oliveira O. N. and Janegitz B. C. (2018). Ultralow cost electrochemical sensor made of potato starch and carbon black nanoballs to detect tetracycline in waters and milk. *Electroanalysis*, **30**, 2153–2159.

Deziel N. (2014). Pharmaceuticals in wastewater treatment plant effluent waters. *Scholarly Horizons: University of Minnesota, Morris Undergraduate Journal*, **1**(2), Article 12.

Dhanalakshmi N., Priya T., Thennarasu S., Sivanesan S. and Thinakaran N. (2020). Synthesis and electrochemical properties of environmental free l-glutathione grafted graphene oxide/ZnO nanocomposite for highly selective piroxicam sensing. *Journal of Pharmaceutical Analysis*, **11**(1), 48–56.

Durand M. J., Thouand G., Dancheva-Ivanova T., Vachon P. and Du Bow M. (2003). Specific detection of organotin compounds with a recombinant luminescent bacterium. *Chemosphere*, **52**, 103–111.

El-Ogri F., Ouazzani N., Boraâm F. and Mandi L. (2016). A survey of wastewaters generated by a hospital in Marrakech city and their characterization. *Desalination and Water Treatment*, **57**(36), 1–14.

Environmental Protection Agency (2010). Pharmaceuticals and Personal Care Products (PPCPs): Frequent Questions. U.S. Environmental Protection Agency. N.p., 28 Oct 2010.

Gayen P. and Chaplin B. P. (2016). Selective electrochemical detection of ciprofloxacin with a porous nafion/multiwalled carbon nanotube composite film electrode. *ACS Applied Materials & Interfaces*, **8**, 1615–1626.

Gilbert N. (2012). Drug-pollution law all washed up. *Nature*, **491**, 503–504.

González-Martínez M. A., Puchades R., Maquieira A., Ferrer I., Marco M. P. and Barceló D. (1999). Reversible immunosensor for the automatic determination of atrazine. Selection of performance of three polyclonal antisera. *Analytica Chimica Acta*, **386**, 201–210.

Govindhan M., Adhikaria B.-R. and Chen A. (2014). Nanomaterials-based electrochemical detection of chemical contaminants. *RSC Advances*, **4**, 63741–63760.

Guo X. (2013). Single-molecule electrical biosensors based on single-walled carbon nanotubes. *Advanced Materials*, **25**, 3397–3408.

Hu L., Cheng Q., Chen D., Ma M. and Wu K. (2015). Liquid-phase exfoliated graphene as highly-sensitive sensor for simultaneous determination of endocrine disruptors: diethylstilbestrol and estradiol. *Journal of Hazardous Materials*, **283**, 157–163.

Hung L. S. (2003). The SARS epidemic in Hong Kong: what lessons have we learned? *Journal of the Royal Society of Medicine*, **96**, 374–378.

Hutter E. and Maysinger D. (2013). Gold-nanoparticle-based biosensors for detection of enzyme activity. *Trends in Pharmacological Sciences*, **34**, 497–507.

Kümmerer K. (2009). Antibiotics in the aquatic environment–a review–part I. *Chemosphere*, **75**, 417–434.

Leprat P. (1998). Les rejets liquides hospitaliers, quels agents et quelles solutions techniques? *Revue Techniques Hospitalières*, **632**, 49–52.

Leung W. K., To K. F., Chan P. K., Chan H. L., Wu A. K., Lee N., Yuen K. Y. and Sung J. J. (2003). Enteric involvement of severe acute respiratory syndrome-associated coronavirus infection. *Gastroenterology*, **125**, 1011–1017.

Li Y., Zhao X., Li P., Huang Y., Wang J. and Zhang J. (2015). Highly sensitive Fe_3O_4 nanobeads/graphene-based molecularly imprinted electrochemical sensor for 17β-estradiol in water. *Analytica Chimica Acta*, **884**, 106–113.

Ling Y., Xu S. B., Lin Y. X., Tian D., Zhu Z. Q., Dai F. H. and Lu H. Z. (2020). Persistence and clearance of viral RNA in 2019 novel coronavirus disease rehabilitation patients. *Chinese Medical Journal*, **133**(9), 1039–1043.

Lis H., Stepnowski P. and Caban M. (2021). Static renewal and continuous-flow calibration of two types of passive samplers for the monitoring of pharmaceuticals in wastewater. *Microchemical Journal*, **165**, 106121.

Lodder W. and de Roda Husman A. M. (2020). SARS-CoV-2 in wastewater: potential health risk, but also data source. *The Lancet Gastroenterology & Hepatology*, **5**(6), 533–534.

Luo Y., Guo W., Ngo H. H., Nghiem L. D., Hai F. I., Zhang J., Liang S. and Wang X. C. (2014). A review on the occurrence of micropollutants in the aquatic environment and their fate and removal during wastewater treatment. *Science of The Total Environment*, **473–474**, 619–641.

Maier N. M., Franco P. and Lindner W. (2001). Separation of enantiomers: needs, challenges, perspectives. *Journal of Chromatography A*, **906**, 3–33.

Mesdaghinia A., Naddafi K., Nabizadeh R., Saeedi R. and Zamanzadeh M. (2009). Wastewater characteristics and appropriate method for wastewater management in the hospitals. *Iranian Journal Public Health*, **38**, 34–40.

National Center for Biotechnology Information (2012). The NCGC Pharmaceutical Collection: A comprehensive resource of clinically approved drugs enabling repurposing and chemical genomics. US National Library of Medicine National Institutes of Health. Science Translational Medicine, 27 April 2012.

Nikolaou A., Meric S. and Fatta D. (2007). Occurrence patterns of pharmaceuticals in water and 572 wastewater environments. *Analytical and Bioanalytical Chemistry*, **387**, 1225–1234.

Norman A., Hansen L. H. and Sørensen S. J. (2005). Construction of a ColD cda promoter-based SOS-green fluorescent protein whole-cell biosensor with higher sensitivity toward genotoxic compounds than constructs based on recA, umuDC, or sulA promoters. *Applied and Environmental Microbiology*, **71**, 2338–2346.

Okoth O. K., Yan K., Feng J. and Zhang J. (2018). Label-free photoelectrochemical aptasensing of diclofenac based on gold nanoparticles and graphene-doped CdS. *Sensors and Actuators B: Chemical*, **256**, 334–341.

Ong Y., Thai A., Abdul L., Zein S. H. and Tan S. H. (2010). A review on carbon nanotubes in an environmental protection and green engineering perspective. *Brazilian Journal of Chemical Engineering*, **27**(2), 227–242.

Onnerfjord P., Eremin S. A., Emneus J. and Marko-Varga G. (1998). Fluorescence polarization for immunoreagent characterisation. *Journal of Immunological Methods*, **213**, 31–39.

Patel P. D. (2002). (Bio)sensors for measurement of analytes implicated in food safety: a review. *TrAC Trends in Analytical Chemistry*, **21**, 96 115.

Priyadarshi H., Alam A., Gireesh-Babu P., Das R., Kishore P., Kumar S. and Chaudhari A. (2012). A GFP-based bacterial biosensor with chromosomally integrated sensing cassette for quantitative detection of Hg (II) in environment. *Journal of Environmental Sciences*, **24**, 963–968.

Pundir C. S. and Chauhan N. (2012). Acetylcholinesterase inhibition-based biosensors for pesticide determination: a review. *Analytical Biochemistry*, **429**(1), 19–31.

Rao A. D., Dash S., Babu S. V. and Jain I. (2007). Numerical modeling of cyclone impact on the ocean- A case study of the Orissa Supper Cyclone. *Journal of Coastal Research*, **23**(5), 1245–.

Rogers K. R. (2006). Recent advances in biosensor techniques for environmental monitoring. *Analytica Chimica Acta*, **568**, 222–231.

Samendra P. S., Masaaki K., Charles P. G. and Ian L. P. (2014). Rapid detection technologies for monitoring microorganisms in water. *Biosensors Journal*, **3**, 109.

Sang S., Wang Y., Feng Q., Wei Y., Ji J. and Zhang W. (2015). Progress of new label-free techniques for biosensors: a review. *Critical Reviews in Biotechnology*, **36**(3), 1–17.

Senveli S. U. and Tigli O. (2013). Biosensors in the small scale: methods and technology trends. *IET Nanobiotechnology*, **7**, 7–21.

Silva M. and Cesarino I. (2019). Evaluation of a nanocomposite based on reduced graphene oxide and gold nanoparticles as an electrochemical platform for detection of sulfamethazine. *Journal of Composites Science*, **3**(2), 59.

Soldano C., Mahmood A. and Dujardin E. (2010). Production, properties and potential of grapheme. *Carbon*, **48**, 2127–2150.

Sun J. Z., Peter K. G., Si R. W., Zhai D. D., Liao Z. H., Sun D. Z., Zheng T. and Yong Y. C. (2015). Microbial fuel cell-based biosensors for environmental monitoring: a review. *Water Science and Technology*, **71**, 801–809.

Sun X.-M., Ji Z., Xiong M.-X. and Chen W. (2017). The electrochemical sensor for the determination of tetracycline based on graphene /L-cysteine composite film. *Journal of The Electrochemical Society*, **164**, B107–B112.

Tonnina D., Campanella L., Sammartino M. P. and Visco G. (2002). Integral toxicity test of sea waters by an algal biosensor. *Anal. Chim*, **92**, 477–484.

Torrinha A., Oliveira T. M. B. F., Ribeiro F. W. P., Correia A. N., Lima-Neto P. and Morais S. (2020). Application of nanostructured carbon-based electrochemical (Bio) Sensors for screening of emerging pharmaceutical pollutants in waters and aquatic Species: A review. *Nanomaterials*, **10**, 1268.

Tsakona M., Anagnostopoulou E. and Gidarakos E. (2007). Hospital waste management and toxicity evaluation: A case study. *Waste Management*, **27**, 912–920.

Valentini F., Galache F. L., Tamburri E. and Palleschi G. (2013). Single walled carbon nanotubes/ polypyrrole_GO_x composite films to modify gold microelectrodes for glucose biosensors: study of the extended linearity. *Biosensors & Bioelectronics*, **43**, 75–78.

Verlicchi P., Galletti A. and Masotti L. (2010a). Management of hospital wastewaters: the case of the effluent of a large hospital situated in a small town. *Water Science and Technology*, **61**, 2507–2519.

Verlicchi P., Galletti A., Petrovic M. and Barcelo D. (2010b). Hospital effluents as a source of emerging pollutants: an overview of micropollutants and sustainable treatment options. *Journal of Hydrology*, **389**(3–4), 416–428.

Verlicchi P., Galletti A., Petrovic M. and Barcelo D. (2012). Micro-pollutants in hospital effluent: their fate, risk and treatment options. In: Emerging Organic Contaminants and Human Health. The Handbook of Environmental Chemistry, Vol. **20**, D. Barceló (ed.), Springer, Berlin, Heidelberg, pp. 139–171.

Wang Ch. and Yu Ch. (2012). Detection of chemical pollutants in water using gold nanoparticles as sensors: a review. *Reviews in Analytical Chemistry*, **32**(1), 1–14.

Wang X. W., Li J. S., Jin M., Zhen B., Kong Q. X., Song N., Xiao W., Yin J., Wei W., Wang G. J., Si B., Guo B. Z., Liu C., Ou G., Wang M. N., Fang T., Chao F. and Li J. W. (2005). Study on the resistance of severe acute respiratory syndrome-associated coronavirus. *Journal of Virological Methods*, **126**, 171–177.

Whitcombe M. J., Chianella I., Larcombe L., Piletsky S. A., Noble J., Porter R. and Horgan A. (2011). The rational development of molecularly imprinted polymer-based sensors for protein detection. *Chemical Society Reviews*, **40**(3), 1547–1571.

Wu Y., Guo C., Tang L., Hong Z., Zhou J., Dong X., Yin H., Xiao Q., Tang Y., Qu X. and Kuang L. (2020). Prolonged presence of SARS-CoV-2 viral RNA in faecal samples. *The Lancet Gastroenterology & Hepatology*, **5**(5), 434–435.

Yang W., Thordarson P., Gooding J. J., Ringer S. P. and Braet F. (2007). Carbon nanotubes for biological and biomedical applications. *Nanotechnology*, **18**, 1–12.

Yang Z., Li G., Dai X., Liu G., Li G. and Jie Y. (2020). Three cases of novel coronavirus pneumonia with viral nucleic acids still positive in stool after throat swab detection turned negative. *Chinese Journal of Digestion*, **12**, E002–E002.

Zhang Y., Yao S., Zeng H. and Song H. (2010). Chiral separation of pharmaceuticals by high 569 performance liquid chromatography. *Current Pharmaceutical Analysis*, **6**, 114–130.

Zhang S., Sun W., Xu L. I., Zheng X., Chu X., Tian J., Wu N. and Fan Y. (2012). Identification of the para-nitrophenol catabolic pathway, and characterization of three enzymes involved in the hydroquinone pathway, in Pseudomonas sp. 1-7. *BMC Microbiology*, **12**, 27.

Zhang J., Wang S. and Xue Y. (2020a). Fecal specimen diagnosis 2019 novel coronavirus-infected pneumonia. *Journal of Medical Virology*, **92**(6), 680–682.

Zhang Y., Chen H., Zhu S., Shu C., Wang D., Song J., Song Y., Zhen W., Feng Z., Wu G., Xu J. and Xu W. (2020b). Isolation of 2019-nCoV from a stool specimen of a laboratory-confirmed case of the coronavirus disease 2019 (COVID-19). *China CDC weekly*, **2**, 123–124.

Zhang N., Gong Y., Meng F., Bi Y., Yang P. and Wang F. (2020c). Virus shedding patterns in nasopharyngeal and fecal specimens of COVID-19 patients. *Science China Life Sciences*, https://doi.org/10.1007/s11427-020-1783-9

IWA PUBLISHING

doi: 10.2166/9781789062625_0031

Chapter 3

Occurrence of COVID-19 virus in hospital wastewater: treatment pathways and sustainability aspects

Biswajit Debnath[1,2]*, *Ruchitha K. Birawat*[3], *Khushbu K. Birawat*[4] and *Anisha Modak*[5]

[1]*Chemical Engineering Department, Jadavpur University, Kolkata, India*
[2]*Department of Mathematics, ASTUTE, Aston University, Birmingham, UK*
[3]*Department of Civil and Environmental Engineering, University of Illinois, Urbana Champaign, USA*
[4]*Department of Civil Engineering, Global Academy of Technology, Bengaluru, India*
[5]*Gouridevi Institute of Medical College & Hospital, Durgapur, India*
**Corresponding author. biswajit.debnath.ju@gmail.com*

ABSTRACT

The world is in a chaotic state due to the COVID-19 pandemic, caused by the highly contagious SARS-CoV-2 virus. The virus spreads easily and impacts not only humans but also the environment. Viruses are charged colloidal particles that possess the ability to adsorb on surfaces. Sorptive interaction of the viruses with solids influences their behaviour in aquatic environments, soils, sewage sludge, and so on. There is a surge of viral load in the hospital effluents due to the ongoing situation, which is posing problems for the wastewater treatment plants (WWTPs). There is a dire need to understand the interaction and flow of viruses in the environment to prevent future mishaps. Monitoring will help us achieve this goal and aid in environmental impact assessment. This chapter presents the possible fate of COVID-19 virus in the environment with detection and surveillance efforts recognized globally. The chapter further discusses the possible issues and challenges faced by the WWTP workers. Additionally, technological solutions are also suggested as retrieved from published literature. Sustainability aspects are also discussed in brief. The chapter is expected to be useful for future researchers and policymakers.

Keywords: COVID-19, environmental impact, health risk, hospital wastewater, sustainability

CONTENTS

Editor(s): Nadeem A Khan, Viola Vambol, Sergij Vambol, Nastaran Mozaffari and Niloofar Mozaffari

3.1 INTRODUCTION

The world is amidst chaos dealing with a global pandemic caused by SARS-CoV-2, also known as COVID-19. It has been over a year since the emergence of COVID-19 in China. It primarily affects respiratory tracts, but there have been studies which suggest the presence of viral RNA in the faeces of infected individuals (Kitajima *et al.*, 2020). The SARS-CoV-2 has a nucleotide sequence that is 75–80% similar to that of SARS-CoV, hence the name SARS-CoV-2 (Zhu *et al.*, 2020). COVID-19 is also related to MERS and the classical human CoV strain (Gorbalenya *et al.*, 2020).

Studies indicate that both SARS-CoV-2 and viral RNA are shed in body fluids, such as saliva, sputum and faeces excreted by the affected individuals. The virus is accompanied by gastrointestinal symptoms such as diarrhoea along with respiratory symptoms. The quantitative polymerase chain reaction (qPCR) method was selected to determine the concentration of viral RNA in faeces, urine or anal/rectal swabs. It has been reported that the shedding of SARS-CoV-2 RNA continues until 7 weeks after the symptoms' onset (Jiehao *et al.*, 2020; Wu *et al.*, 2020c; Xiao *et al.*, 2020). There are specifically two studies which report on the presence of SARS-CoV-2 in faecal samples (Wang *et al.*, 2020b; Zhang *et al.*, 2020a).

The major transmission route is via droplets but the possibility of hand-to-oral contamination cannot be disregarded as the virus can survive for longer durations on inanimate objects. Previously, there was no concrete evidence available as to the presence of COVID-19 in wastewater. However, Wong *et al.* (2013) made the very first detection of SARS-CoV-2 in wastewater. As the majority of the viral load would be discharged in the hospital effluent, there is a need to formulate solutions to monitor and treat the wastewater to curb the viral load (Vambol *et al.*, 2021). This also calls for monitoring of wastewater as it might become a major source of contamination (La Rosa *et al.*, 2020). Wastewater surveillance helps in epidemic monitoring, and can also help in predicting the spread of the virus and act as an alarm or warning system.

3.2 METHODOLOGY

This article is composed using information gathered from various published research papers, reports, news articles and other verified online sources that provide knowledge on the recent advancement in the field of COVID-19 wastewater treatment. Different online databases were explored, such as ScienceDirect, Springer, Wiley, Google Scholar, arXiv, ResearchGate and so on., in identifying relevant published articles. Searching for keywords and/or combinations of keywords was employed for this purpose – 'COVID-19', 'wastewater', 'hospital wastewater', 'environmental impact', 'treatment technology', 'COVID-19 surveillance', 'plant worker', 'health issues' and so on. The accumulated resources were explored and further segregated to identify relevant literature that focuses on presence of the COVID-19 virus in hospital wastewater. The segregated literature was further examined for identification of possible environmental impacts, available treatment technologies, COVID-19 surveillance, and health issues posed to the wastewater treatment plant (WWTP) workers. Further, a qualitative sustainability analysis was conducted on the existing system dealing with COVID-19 wastewater. The references cited in each relevant literature and in this paper were reviewed and referred properly.

3.3 TRANSMISSION ROUTES OF COVID-19 VIRUS THROUGH WWTPS

The novel SARS-CoV-2 virus is of zoonotic origin and the reservoir species and intermediate host identities are unclear and controversial. Preliminary studies originating from Wuhan, China revealed that aerosol transmission is the primary mode of virus transmission, valid for both symptomatic and asymptomatic patients (Li *et al.*, 2020; Liu *et al.*, 2020). Another mode is fomite transmission which is really dangerous as the pandemic-causing virus can survive longer on metal and plastic surfaces compared to air (Kampf *et al.*, 2020). This increases the risk of nosocomial infections among healthcare workers (Wang *et al.*, 2020a, 2020b).

Virus RNA was identified in both symptomatic and asymptomatic patients' saliva samples which indicates the probability of infection in the salivary gland and further transmission (Gu *et al.*, 2020). The virus was also identified in the faeces and urine samples of patients. The presence of the virus was also found up to 21 days or more after the symptoms diminished (Zhang *et al.*, 2020e). To date there have been several reports that confirm the presence of viral RNA in surface water as well as raw sewage, there is no strong evidence that interlinks the contamination of SARS-CoV-2 in wastewater (Ahmed *et al.*, 2020a, 2020b; Wu *et al.*, 2020a, 2020b, 2020c).

The overall transmission routes can be divided into three categories, namely i) droplet transmission; ii) fomite or inanimate surfaces and iii) faecal-oral route (Gormley *et al.*, 2020; Mohapatra *et al.*, 2020). The primary mode of transmission for SARS-CoV-2 virus is droplet transmission. It is well documented for several other viruses that during activities such as coughing, sneezing, deep breath outs, toilet flushes and sewage treatment and so on. aerosols or droplets are often generated which are infected with viral particles (Chattopadhyay & Taft, 2018). The aerosol particles produced in the WWTPs are usually more sluggish, hence they cannot travel longer distances. On the contrary, small aerosols generated during secondary treatment of sewage may travel longer distances (Chattopadhyay & Taft, 2018). However, the WWTP workers have a higher risk of being exposed to infectious aerosols generated in the plant during the treatment process (Wigginton *et al.*, 2015).

Fomite transmission is very dangerous and requires special attention. As reported by Kampf *et al.* (2020) corona virus can survive on plastic surfaces for up to 9 days and on stainless steel for up to 7 days. The World Health Organization (WHO) has released guidelines for handling of SARS-CoV-2 samples which take into account the specific transmissibility of the virus through inanimate surfaces (WHO, 2020a, 2020b, 2020c). Infectivity of corona virus can effectively be diminished within 60 seconds, or less, using disinfectants containing 62–71% ethanol, 0.5% hydrogen peroxide, and 0.1% sodium hypochlorite. An outstanding reduction of viral load in the ambient air and inanimate surfaces has been observed with the implementation of regular disinfection-cum-sanitization of hospitals, isolation wards, anterooms and toilets in Singapore and China (Ong *et al.*, 2020; Wang *et al.*, 2020a, 2020b, 2020e).

During a pandemic, water bodies are highly susceptible to viral contamination which can happen via faeces of patients (Mohapatra *et al.*, 2020; Yeo *et al.*, 2020). Several complications, including gastrointestinal and respiratory issues, can arise due to virus-contaminated faecal matter. It has also been found that in developing countries, faecal matter with high viral load can enter water bodies either through leakage of sewers (Majumdar *et al.*, 2019; Collivignarelli *et al.*, 2020). For instance, disease transmission among family members can occur during toilet flushes by aerosolization of the corona virus (McDermott *et al.*, 2020). Disease transmission can occur due to substandard sanitization of surfaces. These surfaces can contaminate the host's hands or other body parts, ultimately leading to further spread of the virus. Contamination can also happen through vomits and greywater generated during the regular activities of the patients. In China, out of 1099 COVID-19 positive patients from 552 hospitals, nearly 4% of the patients developed diarrhoea while 5% of the patients presented vomiting and/or nausea symptoms (Hindson, 2020). Additionally, almost 49% of the COVID-19-affected people visited the hospitals due to digestive symptoms only (Hindson, 2020). Potentially infectable virus was detected via electron microscopy in faecal samples of COVID-19 positive patients who did not exhibit diarrhoea as a major symptom (Hindson, 2020). Aerosolization from faeces and urine has led to a high concentration of SARS-CoV-2 virus inside the inpatient toilets in a hospital in Wuhan (Liu *et al.*, 2020).

SARS-CoV-2 viruses have a lipid envelope which allows them to survive in strong alkaline environments and lower temperatures (<278 K). As a result, they are more likely to retain their activity in hospital wastewater and greywater for a significant amount of time, which has already been reported for the SARS-CoV virus (Wang *et al.*, 2005b). Hence, the increasing cases of diarrhoea can be linked to virus transmission through wastewater management systems and it needs additional attention for critical evaluation. Another important observation is that corona viruses have a high survival rate in human faeces (Yeo *et al.*, 2020). Consideration of both of these findings, suggests that faecal transmission of these viruses is a possibility. Very recently, some review articles have detailed possible frameworks of COVID-19 viruses in the faecal-oral transmission route (Foladori *et al.*, 2020; Heller *et al.*, 2020). It is only a matter of time before this may be validated empirically. Researchers have already developed models that dictate the transmission routes and results on the environment. Based on the assumption that COVID-19 virus is transmissible through sewage into natural water bodies, Franklin and Bevins (2020) developed a conceptual model that describes transmission of SARS-CoV-2 virus from WWTPs into the natural water bodies, infecting wild animals and human beings; which implies that the virus can also infect WWTP workers. However, these models can provide more precise results if additional factors are taken into account including environmental factors, population density, persistence, infectivity and aerosolization of virus (Mohapatra *et al.*, 2020).

3.4 IDENTIFICATION AND PANDEMIC SURVEILLANCE THROUGH WWT IN DIFFERENT COUNTRIES

Accepting and grasping the full situation which COVID-19 has left us in, is an almost unending challenge for public-health executives, and also all epidemiological indicators have biases and boundaries. Other viral diseases have resulted in the development of accurate monitoring of sewage for traces of a pathogens, thereby enabling efficient inspection of complete communities, which eventually provides the suspected whereabouts of the pathogen, whether it is present in the population and whether transmission is mounting or fading. Researchers are now globally chasing a similar strategy for COVID-19 with the optimism that wastewater data can boost existing measures of its prevalence. SARS-CoV-2, also known as coronavirus, has already been detected in wastewater (Saba *et al.*, 2021). There are several benefits of surveillance of wastewater for COVID-19, including that the process is an economic way to study the transmission kinetics of the virus in the community. In the early stage of the prevailing pandemic, the cumulative cases of COVID-19-diagnosed patients was a primary indicator. Whereas, in the later stage, more recently, indicators such as hospitalization rates, number of deaths, serological data and so on. have been given more importance. These indicators are useful but they have biases which are absent in wastewater surveillance (Saba *et al.*, 2021). Asymptomatic civilians found in community areas such as malls, airports and markets, and symptomatic patients' waste from hospitals are the biggest source of virus shedding in municipal wastewater. Additionally, wastewater surveillance not only removes the possibility of different biases that arise from typical epidemiological indicators but also facilitates data collection from unprivileged patients who lack proper healthcare access (Saba *et al.*, 2021). If wastewater surveillance excels in revealing the transmission dynamics before diagnostic results, then it can present nearly real-time information on disease prevalence to the concerned stakeholders (Larsen & Wigginton, 2020). Table 3.1 provides such information.

SARS-CoV-2 virus can stay alive in sewer networks if the waste/wastewater remains untreated, which could become a secondary origin of infection spreading (Saba *et al.*, 2021; Zhang *et al.*, 2020b). According to Saba *et al.* (2021), monitoring of SARS-CoV-2 in wastewater treatment plants and

Table 3.1 Pandemic surveillance through wastewater samples.

Sl. No.	Country	Source of Sample	Concentration of SARS-CoV–2 Virus (genome copies/L)	References
1.	India	Untreated wastewater	$0.56–3.5 \times 10^2$	Kumar *et al.* (2020)
2.	China	Hospital wastewater	$0.5–18.7 \times 10^3$	Zhang *et al.* (2020c)
3.	Spain	Wastewater treatment plants	2.5×10^2	Randazzo *et al.* (2020)
4.	Brazil	Wastewater treatment plants	$0.103–13.1 \times 10^3$	Zaneti *et al.* (2021)
5.	France	Wastewater treatment plants	50.3×10^3	Wurtzer *et al.* (2020)
6.	Turkey	Primary and waste activated sludge	$1.17–4.02 \times 10^4$	Kocamemi *et al.* (2020)
7.	Japan	Biologically treated wastewater (before chlorination)	$1.2–25 \times 10^2$	Haramoto *et al.* (2020)
8.	Australia	Untreated wastewater & wastewater treatment plants	1.9–12	Ahmed *et al.* (2020a)
9.	UK	Municipal wastewater	1.5×10^4	Hillary *et al.* (2021)
10.	Italy	Wastewater pumping station	11.04–11.27	Baldovin *et al.* (2021)

decentralized WWTPs is important and should be treated as an emergency step towards pandemic monitoring. A case study in New Haven, Connecticut, USA reveals that measuring the concentrations of SARS-CoV-2 RNA in primary sewage sludge led to an understanding of the infection dynamics. The study was also able to provide 1–4 days of advance notice period before hospitalization (Peccia *et al.*, 2020). Detection of SARS-CoV-2 virus in municipal wastewater via regular specimen collection can generate a huge amount of relevant data. These can be treated as equivalent to random testing of numerous civilians, but is a less expensive and invasive method. Wastewater data could also be utilized to verify the trustworthiness of epidemiological trends determined from diagnosed cases. There is a pressing necessity to develop wastewater-based epidemiology (WBE) techniques for SARS-CoV-2 in wastewater to achieve early detection in a community and thereby protection of community health by implementing preventive measures (Daughton, 2020). It is essential to create an assessment to define the scale of infection spread and to identify sensors for swift WBE and sharing of data between states and countries. Brief details of WBE in different countries is provided in Table 3.1.

3.4.1 Success story 1: England

Wastewater based epidemiology has been proved successful when applied to infectious diseases, and markers of human health and behaviour in wastewaters have been established in many countries for monitoring of SARS-CoV-2, for instance in England (Baraniuk, 2020; Hillary *et al.*, 2021). A WBE success story of England has been reported by Morvan *et al.* (2021). Data were collected and analysed between July 2020 and March 2021 from 44 sewage treatment works (STWs) across England which cover an estimated 31% of the population. For every site, an average of four samples were collected per week and were eventually analysed, by either grab (41%) or composite (59%) sampling. Site and sample level variables were combined with wastewater data and were applied within a statistical modelling framework to estimate the prevalence of SARS-CoV-2. As a result, reliable metrics were provided across regions and throughout the evolving epidemiology of the COVID-19 pandemic in England (Morvan *et al.*, 2021).

The period when these wastewater data were collected in England marked the prevalence of infection which varied considerably as a consequence of the emergence of the epidemic and suppression through non-pharmaceutical interventions. Statistical modelling was presented which portrayed that estimation of prevalence was accurate and precise for a wide range of values. As per the model results, SARS-CoV-2 prevalence was tracked within 1.1% (with 95% confidence) of the Coronavirus infection survey. Best-fitting models were applied to observe a 4-day lead in wastewater concentration over mainly symptomatic testing through routine surveillance (Morvan *et al.*, 2021). Metagenomic analysis of wastewater samples provides insight into the genomic diversity of viruses in the community. The sampling of sites that cover small catchment areas remains capable of revealing localized spikes to detect hotspots and eventually inform the local public health authorities (Office for National Statistics [ONS], 2021). Later on, the investment in wastewater surveillance systems across England, combined with insights from analysis pipelines were applied to other communicable and non-communicable diseases and also to human behaviours which, in turn, impact personal and community health (Hassard *et al.*, 2021; Morvan *et al.*, 2021; Polo *et al.*, 2020).

3.5 DISCUSSION AND ANALYSIS

3.5.1 Issues and challenges

Although the presence of COVID-19 virus in hospital wastewater provides an opportunity for early detection and epidemic monitoring, this comes with caveats. Firstly, untreated hospital wastewater might become a substantial source of contamination. Thus, control, neutralization and treatment of untreated hospital wastewater is an essential task (Bogler *et al.*, 2020). Secondly, during the monitoring and treatment stage the personnel involved might be susceptible to infection and, thus, precautionary protocols have to be followed stringently.

In order to detect and monitor SARS-CoV-2 virus in wastewater, we need advance technology that can create sufficient concentration of viral particles in water samples for accurate quantification (Achak *et al.*, 2021). Lu *et al.* (2020) compiled and presented several approaches for effective primary concentration of SARS-CoV-2 viral particles from wastewater.

There are three main approaches for the detection and monitoring of SARS-CoV-2 in wastewater, namely, qualitative, quantitative molecular, and *in vitro* counts by plaque-forming units (PFU) (Bogler *et al.*, 2020). These approaches vary as they have different capabilities for detection, different speeds of detection, and different levels of sensitivity, which need certain virus concentrations to be present in wastewater.

In order to understand the major environmental pathways and fate of the virus, it is important to identify the feasibility of testing and detection methodology for a larger-scale implementation. This means, we need to have a substantial network of testing but with an economical approach. Studies have been conducted at community level, to identify the transmission of the virus, using a WBE approach (Daughton, 2020). In a study conducted by Hart and Halden (2020), it was found that one infected individual is detectable among 100–2 000 000 persons through wastewater monitoring. This would mean 2.1 billion people could be monitored globally in 105 600 sewage treatment plants, potentially saving billions of dollars.

Viruses have relatively higher resistance to conventional water disinfection processes compared to bacteria (García-Gil *et al.*, 2020). Thus, there is a need to search for novel, cost-effective methodologies for viral inactivation in water (Bandala *et al.*, 2021). Tertiary treatment processes have been reported to be effective for virus inactivation with varying efficiency (Carducci *et al.*, 2020), and several different technologies have been tested for the inactivation of different viruses in water including ozone (O_3) deactivation, solar disinfection, wet peroxide oxidation, chlorine disinfection, cold plasma, electrocoagulation-electro oxidation, photocatalysis with TiO_2, graphene oxide (GO), engineered nanomaterials and so on. (Bandala *et al.*, 2021). Advanced oxidation processes (AOPs) have great potential as they have been tested at bench scale whereas ozone and solar disinfection have already had field application (Heffron *et al.*, 2019; Zhang *et al.*, 2020c, 2020d). However, the number of engineered nanomaterials for virus inactivation are quite few compared to the plethora of different nanomaterials available (Bandala & Berli, 2019).

Although there is sufficient evidence that hospital wastewater can be used for epidemic monitoring (Larsen & Wigginton, 2020; Peccia *et al.*, 2020), the concentration of virus present in the wastewater is very small. Additionally, it is a very time consuming process. Thus, the current technology available needs to be developed further to test for the presence of COVID-19, even when it is present in very low concentrations. Research in this direction is in progress (Gonçalves *et al.*, 2021).

3.5.2 Health risk of WWTP workers

WWTP workers may be exposed to many microorganisms, including SARS-CoV-2, when carrying out routine activities like sewage sampling, manual screening of coarse cleansing, and chemical analysis, especially in regions of poor sanitation infrastructure and with improper personal protective equipment (PPE) (Zaneti *et al.*, 2021). Figure 3.1 depicts the possible pathways of health hazard for the WWTP workers which include both primary contamination from wastewater and sludge handling as well as transmission from asymptomatic fellow workers (Amoah *et al.*, 2020; Dada & Gyawali, 2021; Zaneti *et al.*, 2021). Detection of genetic material nucleic acid (RNA) in the wastewater system has been a supplemental tool for early tracking of SARS-CoV-2 infections (Ahmed *et al.*, 2020a, 2020b; Medema *et al.*, 2020; Wu *et al.*, 2020a, 2020b, 2020c). Qualitative microbial risk assessment (QMRA) has been an essential tool for detecting and confirming estimation of infectivity among workers associated with aerosols or droplets inhalation, though transmission through the faecal-oral route remains a debated issue (Haminton and Haas 2016). Aerosolized viruses can be transported locally by wind at the time of irrigation, during fertilization and via water bodies for agricultural purposes (Gholipour *et al.*, 2021).

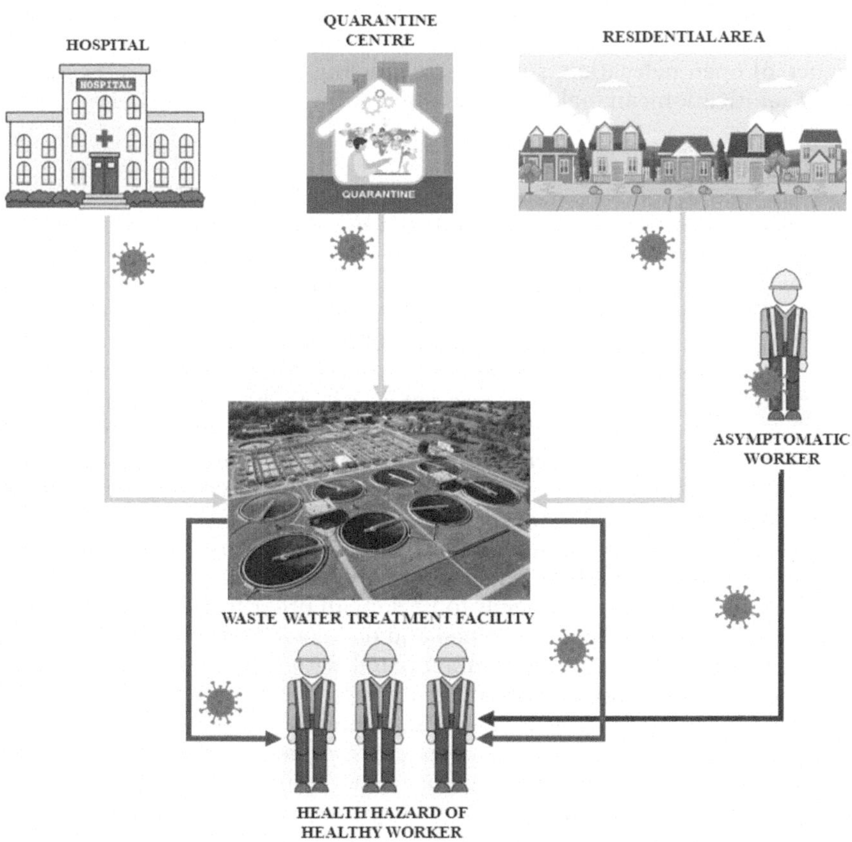

Figure 3.1 Possible pathways of health hazard for WWTP workers.

Workers could be exposed to pathogens when working with class B biosolids. (Brisolara *et al.*, 2021). In this unprecedented situation, the provision of wastewater collection has become a prodigious challenge for workers as has maintaining hygiene and disinfection of reusable PPE (Gormley *et al.*, 2020). In industrialized countries, collection of domestic wastewater and viral load treatment is conducted conventionally which generally does not completely remove virions, which can lead to creation of potential transmission routes during pandemics (Wurtzer *et al.*, 2020). SARS-CoV-2 primarily causes respiratory syndrome but can infect the gastrointestinal tract manifesting in symptoms like diarrhoea and vomiting, resulting in the virus entering the wastewater system through human secretions (urine, stool, sputum) (La Rosa *et al.*, 2020). SARS-CoV-2 can remain viable on waste materials such as plastic and other hard surfaces which may then become a source of infection (Long *et al.*, 2020). Furthermore, survival time of enveloped SARS-CoV-2 depends on numerous environmental factors, such as temperature, pH of solution and organic content, and may remain infective after exposure to wastewater systems, during climate change, or due to human activities (La Rosa *et al.*, 2020; Chen *et al.*, 2020). Inappropriate hand washing after contacting with biosolids, the practice of open defecation, smoking or chewing tobacco while working, and lack of sanitization can make workers more vulnerable to COVID-19.

3.5.3 Technological solutions

The viral load in the wastewater can be detected by concentrating the virus, as a very low quantity of virus is usually present. Concentration can be carried out by electronegative membrane-direct RNA extraction, ultrafiltration, polyethylene glycol (PEG) precipitation, and ultracentrifugation (Adelodun *et al.*, 2020; Ahmed *et al.*, 2020a, 2020b; Medema *et al.*, 2020; Nemudryi *et al.*, 2020; Wu *et al.*, 2020b; Wurtzer *et al.*, 2020). In recent times, quantitative reverse transcription polymerase chain reaction (RT-qPCR) and nested reverse transcription polymerase chain reaction (RT-PCR) have been employed for the detection of the viral load in wastewater samples. A plaque forming test, enzyme-linked immunosorbent assay (ELISA) and biosensors can also be employed for COVID-19 detection and concentration in wastewater. Table 3.2 summarizes details of different treatment technologies.

As hospitals are a major source of SARS-Cov-2 in wastewater, we have to come up with treatment plans which can be applied before the water is released into other water bodies. This needs to be done in order to reduce the spread of the virus. There are physical, chemical and biological methods which can help in the treatment of wastewater contaminated with viruses. Physical methods include use of radiation, gamma rays, UV-rays and heat. Chemical methods employ disinfectants (Kozub *et al.*, 2021). There are various types of chlorine-based disinfectants: hypochlorite, hypochlorous acid, chloramine, chlorine dioxide, chlorine and sodium dichloro isocyanurate. Hydrogen peroxide, quaternary ammonium compounds, and organic peroxides, like peracetic acid and performic acid, can be used as disinfecting agents (Kataki *et al.*, 2021).

Table 3.2 Effectiveness of technological solutions.

Sl No.	Technology	Advantage	Disadvantage	References
1.	Adsorption	• Low cost • High efficiency	• Surfactant recovery • Skilled labour	Valizadeh *et al.* (2016), Teymoorian *et al.* (2021)
2.	Biological treatment	• Low cost • Easy operation	• High residence time • High sludge volume	Deng and Zhao (2015), Teymoorian *et al.* (2021)
3.	Coagulation &flocculation	• Easy operation • High removal efficiency	• Toxic solid compound generation	Teymoorian *et al.* (2021)
4.	Advanced oxidation process	• High reaction rate • Small footprint	• Residual peroxide removal	Oller *et al.* (2011), Teymoorian *et al.* (2021)
5.	UV light disinfection	• Low investment • Cheap operation	• Little penetration • Occupational health hazard	Wang *et al.* (2020a, 2020b, 2020c)
6.	Chlorine dioxide disinfection	• Low-cost operation • High efficiency	• Storage and transportation problem	Wang *et al.* (2020a, 2020b, 2020d)
7.	Biosensors	• Rapid virus detection • Low cost • Easy transport • High sensitivity	• Antibody binding capacity is dependent on other conditions • Antigen-antibody interaction is highly sensitive and easily disrupted by physical and chemical constraints	Lahrich *et al.* (2021)

Viruses have an inherent property to adsorb to a surface. This property is influenced by pH, electrolytes, interfering particles and so on. The rate of adsorption and influencing factors can help us design a protocol for removal of viruses in wastewater. By limiting the carbon and nutrient source, the virus can be inactivated. Bioreactors with specific types of membranes can be used to aid in reduction of viruses. Nanofibres with electric charge have charges which attract the genetic material of the virus (Venugopal *et al.*, 2020). Other techniques like biosorption and photodegradation are also of importance. Algal-based WWTPs are of interest as they are sustainable and cost-effective for virus inactivation by limiting carbon and nutrients (Buchanan, 2014; Young *et al.*, 2016).

The development of hybrid treatment systems can increase the efficiency of viral treatment in wastewater, and the use of robust disinfection techniques can remove viruses as well. Biotechnological tools can be applied to advance and classify aggressive microorganisms that can destroy viruses in wastewater. In this way, the treatment efficiency of prevailing conventional WWTPs is increased. Integration of tertiary treatment in the existing plants will significantly increase the treatment cost. Nevertheless, it is advisable that a single-point investment in clean technologies should be made as it pays off in long-term cost.

New and modern technologies, such as biosensors, are required to be used in wastewater treatments plants to detect and prevent the spread of viruses. There are a number of advantages to using these new biosensors, including faster reaction times and increased detection rates. Biosensor-based microorganism detection in wastewater is an emerging topic in the field of environmental research (Ramírez-Castillo *et al.*, 2015; Yu *et al.*, 2017). There are many different methods available including optical, electrochemical, acoustic, thermal, osmotic and biochemical detection. Compared to traditional methods, biosensors provide real-time identification of microorganisms, accurate quantification and fast response. These features make biosensors especially suitable for wastewater analysis. A wide variety of microorganisms, including bacteria, viruses, fungi and protozoa, can be detected in wastewater using a variety of biosensor technologies. The most common sensors used for this purpose are planar waveguides (PWGs), quartz crystal microbalances (QCMs), optical fibre probes (OFPs) and also nanoparticle-based sensors such as nano-electrodes and carbon nanotube transducers. Currently available biosensors not only enable the detection of multiple virus types in water samples but also detect cultivable organisms such as bacteria which were not detectable in earlier methods. Use of nano-biosensors in wastewater monitoring for water safety and quality has been increasing as they offer higher precision and sensitivity (Ansari, 2017). Viral-protein-detecting biosensors, electrochemical, or optical biosensors could help identify viruses in wastewater samples.

For ameliorating treatment efficiency in WWTPs in the decision-making phase, Artificial Intelligence (AI) is being used. AI is used to forecast, estimate, and diagnose operations in a WWTP. Recently, AI has been applied in four aspects of a WWTP, namely (i) treatment technology; (ii) management; (iii) economy; and (iv) reuse (Zhao *et al.*, 2020). This study revealed that by application of AI, operational cost can be decreased by almost one-third with high precision outputs. Amalgamation of biosensors with AI can be utilized as alarm system which can predict the existence of viruses and thereby allow strategizing of the treatment process accordingly.

3.5.4 Sustainability aspects

Sustainability is a systems approach where the analysis is usually performed within the system boundary (Costanza and Patten, 1995; Dresner, 2012; Wilkinson *et al.*, 2001; Kuhlman & Farrington, 2010). In the current scenario, defining a boundary for sustainability analysis of WWT systems handling wastewater containing SARS-CoV-2 virus shedding is difficult. However, to makes things simple, we use a source-to-sink approach to consider a system boundary, as a generalized case. For example, hospital to WWT facility, where the hospital is the source and the WWT facility is the sink. This imitates a utilization-to-cradle scenario (Kakadellis & Harris, 2020). We also assume the technological solutions provided in the previous section prevail. Given the complexity and the lack of reliable data, we present a few observations from the perspective of sustainability. Here we consider four pillars of sustainability that is environmental, economic, operational and social (Debnath & Ghosh, 2019; Debnath *et al.*, 2014). Our analysis reveals the following points:

(i) The existing technology may not be sufficient to tackle this contumacious virus and hence requires the addition of either advanced monitoring tools or further tertiary/quaternary treatment methods which involve the use of chemicals to destroy the virus. The chemicals need to be neutralized before the final effluent is discharged. This will have an increased environmental impact. Some of the possible impact categories include freshwater ecotoxicity, marine ecotoxicity, terrestrial ecotoxicity, terrestrial acidification, marine eutrophication, freshwater eutrophication, photochemical oxidant formation, water depletion, metal depletion and so on. (Morera *et al.*, 2017; Szulc *et al.*, 2021).

(ii) Utilization of biosensors and associated monitoring technologies will incur extra cost. Similarly, the addition of any further technological solutions will add extra investment cost. In addition, there will be associated operating costs. Electricity consumption will also be increased. This will affect the economic sustainability. In addition to this, higher consumption of electricity increases environmental impact with possible impact categories of global warming potential, fossil fuel depletion and fresh water depletion (Dong & Hauschild, 2017; Szulc *et al.*, 2021).

(iii) Any increase in cost will be offset from the tax payer's money. Hence, this will have social implications. In addition to this, the field workers will have increased safety requirements to stop cross-contamination of the virus. There could be social uncertainty in the supply chain with any minor external perturbation (Chattopadhyay *et al.*, 2020). Given the sensitivity of the factors along the supply chain, socio-economic sustainability is more likely to be affected first. Hence, utmost attention should be paid in these cases.

3.6 CONCLUSION

Wastewater-based epidemiology complements and aggrandizes traditional surveillance, with enormous implications for public health. For infection prevalence, quantification of SARS-CoV-2 RNA in wastewater needs to be executed. In this chapter, we have discussed the transmission routes of SARS-CoV-2 virus through wastewater and how they can be useful in pandemic surveillance. Experiences from different countries have been conglomerated along with a detailed success story of England. The issues and challenges, with a focus on the health risk of the WWTP workers, have been elucidated. In addition, the technological solutions have been detailed. The sustainability aspects have also been provided from a systems perspective. The findings of this chapter will not only help the research community but also the policymakers to a certain extent.

ACKNOWLEDGEMENTS

The authors acknowledge the support received from Jadavpur University, India; Global Academy of Technology, Bengaluru, India and Aston University, UK. The authors also acknowledge sustainability inputs from Ms. Ankita Das, MAKAUT, India and Dr. Saswati Gharami, Jadavpur University, India. The findings, opinions, concepts, implications and/or conclusions articulated in this chapter are solely based on views of the author(s) and do not automatically reflect the opinions of the authors' universities.

AUTHOR CONTRIBUTIONS

The work was conceptualized by BD together with KKB. The primary basis of the work was corroborated by BD, RKB and KKB. Data mining and draft writing was carried out by all the authors. The health risk section was analysed and composed by AM. BD composed the sustainability section with the help of AM and KKB. Proof reading was carried out by KKB and BD. This version of the chapter has been validated by all the authors.

REFERENCES

Achak M., Bakri S. A., Chhiti Y., M'hamdi Alaoui F. E., Barka N. and Boumya W. (2021). SARS-CoV-2 in hospital wastewater during outbreak of COVID-19: A review on detection, survival and disinfection technologies. *Science of the Total Environment*, **761**, 143192.

Adelodun B., Ajibade F. O., Ibrahim R. G., Bakare H. O. and Choi K. S. (2020). Snowballing transmission of COVID-19 (SARS-CoV-2) through wastewater: Any sustainable preventive measures to curtail the scourge in low-income countries? *Science of the Total Environment*, **742**, 140680, https://doi.org/10.1016/j.scitotenv.2020.140680

Adelodun B., Ajibade F. O., Ibrahim R. G., Bakare H. O. and Choi K. S. (2020). Snowballing transmission of COVID-19 (SARS-CoV-2) through wastewater: Any sustainable preventive measures to curtail the scourge in low-income countries? *Science of the Total Environment*, **742**, 140680, https://doi.org/10.1016/j.scitotenv.2020.140680

Ahmed W., Angel N., Edson J., Bibby K., Brien J. W. O., Choi P. M., Kitajima M., Simpson L., Li J., Tscharke B., Verhagen R., Smith J. M., Zaugg J., Dierens L., Hugenholtz P., Thomas K. V. and Mueller J. F. (2020a). First confirmed detection of SARS-CoV-2 in untreated wastewater in Australia: a proof of concept for the wastewater surveillance of COVID-19 in the community. *Science of The Total Environment*, **728**, 138764, https://doi.org/10.1016/j.scitotenv.2020.138764

Ahmed W., Bertsch P. M., Bivins A., Bibby K., Farkas K., Gathercole A., Haramoto E., Gyawali P., Korajkic A., McMinn B. R., Mueller J. F., Simpson S. L., Smith W. J. M., Symonds E. M., Thomas K. V., Verhagen R. and Kitajima M. (2020b). Comparison of virus concentration methods for the RT-qPCR-based recovery of murine hepatitis virus, a surrogate for SARS-CoV-2 from untreated wastewater. *Science of The Total Environment*, **739**, 139960, https://doi.org/10.1016/j.scitotenv.2020.139960

Amoah I. D., Kumari S. and Bux F. (2020). Coronaviruses in wastewater processes: source, fate and potential risks. *Environment International*, **143**, 105962, https://doi.org/10.1016/j.envint.2020.105962

Ansari S. (2017). Combination of molecularly imprinted polymers and carbon nanomaterials as a versatile biosensing tool in sample analysis: Recent applications and challenges. *Trends in Analytical Chemistry*, **93**, 134–151.

Baldovin T., Amoruso I., Fonzo M., Buja A., Baldo V., Cocchio S. and Bertoncello C. (2021). SARS-CoV-2 RNA detection and persistence in wastewater samples: An experimental network for COVID-19 environmental surveillance in Padua, Veneto Region (NE Italy). *Science of The Total Environment*, **760**, 143329, https://doi.org/10.1016/j.scitotenv.2020.143329

Bandala E. R. and Berli M. (2019). Engineered nanomaterials (ENMs) and their role at the nexus of Food, Energy, and Water. *Materials Science for Energy Technologies*, **2**(1), 29–40, https://doi.org/10.1016/j.mset.2018.09.004

Bandala E. R., Kruger B. R., Cesarino I., Leao A. L., Wijesiri B. and Goonetilleke A. (2021). Impacts of COVID-19 pandemic on the wastewater pathway into surface water: A review. *Science of the Total Environment*, **145586**.

Baraniuk C. (2020). Sewage monitoring is the UK's next defence against COVID-19. *BMJ*, **370**, m2599.

Bogler A., Packman A., Furman A., Gross A., Kushmaro A., Ronen A., Dagot C., Hill C., Vaizel-Ohayon D., Morgenroth E. and Bertuzzo E. (2020). Rethinking wastewater risks and monitoring in light of the COVID-19 pandemic. *Nature Sustainability*, **3**(12), 981–990, https://doi.org/10.1038/s41893-020-00605-2

Brisolara, K.F., Maal-Bared, R., Sobsey, M.D., Reimers, R.S., Rubin, A., Bastian, R.K., Gerba, C., Smith, J.E., Bibby, K., Kester, G. and Brown, S. (2021). Assessing and managing SARS-CoV-2 occupational health risk to workers handling residuals and biosolids. *Science of the Total Environment*, **774**, 145732, https://doi.org/10.1016/j.scitotenv.2021.145732

Buchanan A. N. (2014). Comparing the performance of a high rate algal pond with a waste stabilisation pond in rural South Australia (Doctoral dissertation, Flinders University, School of the Environment).

Carducci A., Federigi I., Liu D., Thompson J. R. and Verani M. (2020). Making waves: coronavirus detection, presence and persistence in the water environment: state of the art and knowledge needs for public health. *Water Research*, **179**, 115907, https://doi.org/10.1016/j.watres.2020.115907

Chattopadhyay S. and Taft S. (2018). Exposure Pathways to High-consequence Pathogens in the Wastewater Collection and Treatment Systems. U.S. Environmental Protection Agency, EPA/600/R-18/221.

Chattopadhyay A. K., Debnath B., El-Hassani R., Ghosh S. K. and Baidya R. (2020). Cleaner Production in Optimized Multivariate Networks: Operations Management through a Roll of Dice. arXiv preprint arXiv:2003.00884.

Chen N., Zhou M., Dong X., Qu J., Gong F., Han Y., Qiu Y., Wang J., Liu Y., Wei, Y., Xia, J., Yu, T., Zhang, X., and Zhang, Li. (2020). Epidemiological and clinical characteristics of 99 cases of 2019 novel coronavirus pneumonia in Wuhan, China: a descriptive study. *The Lancet*, **395**(10223), 507–513.

Collivignarelli M. C., Collivignarelli C., Miino M. C., Abbà A., Pedrazzani R. and Bertanza G. (2020). SARS-CoV-2 in sewer systems and connected facilities. Process safety and environmental protection.

Costanza R. and Patten B. C. (1995). Defining and predicting sustainability. *Ecological Economics*, **15**(3), 193–196, https://doi.org/10.1016/0921-8009(95)00048-8

Dada A. C. and Gyawali P. (2021). Quantitative microbial risk assessment (QMRA) of occupational exposure to SARS-CoV-2 in wastewater treatment plants. *Science of The Total Environment*, **763**, 142989, https://doi.org/10.1016/j.scitotenv.2020.142989

Daughton C. G. (2020). Wastewater surveillance for population-wide COVID-19: the present and future. *Science of the Total Environment*, **736**, 139631, https://doi.org/10.1016/j.scitotenv.2020.139631

Debnath B. and Ghosh S. K. (2019). Hydrogen generation from biorefinery waste: recent advancements and sustainability perspectives. In: Waste Valorisation and Recycling. Springer, Singapore, pp. 557–572.

Debnath B., Biswas N. T., Baidya R. and Ghosh S. K. (2014). Nanotechnology in waste water treatment: a review. *Ecology of Urban Areas*, 563–575.

Deng Y. and Zhao R. (2015). Advanced oxidation processes (AOPs) in wastewater treatment. *Current Pollution Reports*, **1**(3), 167–176, https://doi.org/10.1007/s40726-015-0015-z

Dong Y. and Hauschild M. Z. (2017). Indicators for environmental sustainability. *Procedia Cirp*, **61**, 697–702, https://doi.org/10.1016/j.procir.2016.11.173

Dresner S. (2012). The Principles of Sustainability. Routledge, London.

Foladori P., Cutrupi F., Segata N., Manara S., Pinto F., Malpei F., Bruni L. and La Rosa G. (2020). SARS-CoV-2 from faeces to wastewater treatment: what do we know? A review. *Science of the Total Environment*, **743**, 140444, https://doi.org/10.1016/j.scitotenv.2020.140444

Franklin A. B. and Bevins S. N. (2020). Spillover of SARS-CoV-2 into novel wild hosts in North America: A conceptual model for perpetuation of the pathogen. *Science of the Total Environment*, **733**, 139358, https://doi.org/10.1016/j.scitotenv.2020.139358

García-Gil Á, Martínez A., Polo-López M. I. and Marugán J. (2020). Kinetic modeling of the synergistic thermal and spectral actions on the inactivation of viruses in water by sunlight. *Water Research*, **183**, 116074, https://doi.org/10.1016/j.watres.2020.116074

Gholipour, S., Mohammadi, F., Nikaeen, M., Shamsizadeh, Z., Khazeni, A., Sahbaei, Z., Mousavi, S.M., Ghobadian, M. and Mirhendi, H. (2021). COVID-19 infection risk from exposure to aerosols of wastewater treatment plants. *Chemosphere*, **273**, 129701, https://doi.org/10.1016/j.chemosphere.2021.129701

Gonçalves, J., Koritnik, T., Mioč, V., Trkov, M., Bolješič, M., Berginc, N., Prosenc, K., Kotar, T. and Paragi, M., (2021). Detection of SARS-CoV-2 RNA in hospital wastewater from a low COVID-19 disease prevalence area. *Science of The Total Environment*, **755**(Pt 2), 143226, https://doi.org/10.1016/j.scitotenv.2020.143226

Gorbalenya, A.E., Baker, S.C., Baric, R.S., de Groot, R.J., Drosten, C., Gulyaeva, A.A., Haagmans, B.L., Lauber, C., Leontovich, A.M., Neuman, B.W. and Penzar, D., (2020). Coronaviridae Study Group of the International Committee on Taxonomy of Viruses. The species severe acute respiratory syndrome-related coronavirus: classifying 2019–nCoV and naming it SARS-CoV-2. *Nature Microbiology*, **5**(4), 536–544, https://doi.org/10.1038/s41564-020-0695-z

Gormley M., Aspray T. J. and Kelly D. A. (2020). COVID-19: mitigating transmission via wastewater plumbing systems. *The Lancet Global Health*, **8**(5), e643, https://doi.org/10.1016/S2214-109X(20)30112-1

Gu J., Han B. and Wang J. (2020). COVID-19: gastrointestinal manifestations and potential fecal–oral transmission. *Gastroenterology*, **158**(6), 1518–1519, https://doi.org/10.1053/j.gastro.2020.02.054

Hamilton K. A. and Haas C. N. (2016). Critical review of mathematical approaches for quantitative microbial risk assessment (QMRA) of Legionella in engineered water systems: research gaps and a new framework. *Environmental Science: Water Research & Technology*, **2**(4), 599–613.

Haramoto E., Malla B., Thakali O. and Kitajima M. (2020). First environmental surveillance for the presence of SARS-CoV-2 RNA in wastewater and river water in Japan. *Science of The Total Environment*, **737**, 140405, https://doi.org/10.1016/j.scitotenv.2020.140405

Hart O. E. and Halden R. U. (2020). Computational analysis of SARS-CoV-2/COVID-19 surveillance by wastewater-based epidemiology locally and globally: feasibility, economy, opportunities and challenges. *Science of the Total Environment*, **730**, 138875, https://doi.org/10.1016/j.scitotenv.2020.138875

Hassard F., Lundy L., Singer A. C., Grimsley J. and Di Cesare M. (2021). Innovation in wastewater near-source tracking for rapid identification of COVID-19 in schools. *The Lancet Microbe*, **2**(1), e4–e5, https://doi.org/10.1016/S2666-5247(20)30193-2

Heffron J., Ryan D. R. and Mayer B. K. (2019). Sequential electrocoagulation-electrooxidation for virus mitigation in drinking water. *Water Research*, **160**, 435–444, https://doi.org/10.1016/j.watres.2019.05.078

Heller L., Mota C. R. and Greco D. B. (2020). COVID-19 faecal-oral transmission: Are we asking the right questions? *Science of the Total Environment*, **729**, 138919, https://doi.org/10.1016/j.scitotenv.2020.138919

Hillary L. S., Farkas K., Maher K. H., Lucaci A., Thorpe J., Distaso M. A., Gaze W. H., Paterson S., Burke T., Connor T. R. and Jones D. L. (2021). Monitoring SARS-CoV-2 in municipal wastewater to evaluate the success of lockdown measures for controlling COVID-19 in the UK. *Water Research*, **200**, 117214, https://doi.org/10.1016/j.watres.2021.117214

Hindson J. (2020). COVID-19: faecal–oral transmission? *Nature Reviews Gastroenterology & Hepatology*, **17**(5), 259–259, https://doi.org/10.1038/s41575-020-0295-7

Jiehao, C., Jin, X., Daojiong, L., Zhi, Y., Lei, X., Zhenghai, Q., Yuehua, Z., Hua, Z., Ran, J., Pengcheng, L. and Xiangshi, W., (2020). A case series of children with 2019 novel coronavirus infection: clinical and epidemiological features. *Clinical Infectious Diseases*, **71**(6), 1547–1551, https://doi.org/10.1093/cid/ciaa198

Kakadellis S. and Harris Z. M. (2020). Don't scrap the waste: The need for broader system boundaries in bioplastic food packaging life-cycle assessment–A critical review. *Journal of Cleaner Production*, **274**, 122831, https://doi.org/10.1016/j.jclepro.2020.122831

Kampf G., Todt D., Pfaender S. and Steinmann E. (2020). Persistence of coronaviruses on inanimate surfaces and their inactivation with biocidal agents. *Journal of Hospital Infection*, **104**(3), 246–251, https://doi.org/10.1016/j.jhin.2020.01.022

Kataki S., Chatterjee S., Vairale M. G., Sharma S. and Dwivedi S. K. (2021). Concerns and strategies for wastewater treatment during COVID-19 pandemic to stop plausible transmission. *Resources, Conservation and Recycling*, **164**, 105156. https://doi.org/10.1016/j.resconrec.2020.105156

Kitajima, M., Ahmed, W., Bibby, K., Carducci, A., Gerba, C.P., Hamilton, K.A., Haramoto, E. and Rose, J.B. (2020). SARS-CoV-2 in wastewater: state of the knowledge and research needs. *Science of The Total Environment*, **739**, 139076. https://doi.org/10.1016/j.scitotenv.2020.139076

Kocamemi B. A., Kurt H., Sait A., Sarac F., Saatci A. M. and Pakdemirli B. (2020). SARSCoV-2 detection in istanbul wastewater treatment plant sludges. *medRxiv*, **143**, 196–203. https://www.medrxiv.org/content/10.1101/2020.05.12.20099358v1 (accessed 28 October 2021)

Kozub, P., Kozub, S., Mozaffari, N., Sihag, P., Mozaffari, N., Vambol, S., Vambol, V., Ramsawak, N. and Khan, N.A., (2021). Treatment schemes–conventional and dedicated for PhACs treatment. *Pharmaceutical Wastewater Treatment Technologies*, 181–220.

Kuhlman T. and Farrington J. (2010). What is sustainability? *Sustainability*, **2**(11), 3436–3448, https://doi.org/10.3390/su2113436

Kumar M., Patel A. K., Shah A. V., Raval J., Rajpara N., Joshi M. and Joshi C. G. (2020). First proof of the capability of wastewater surveillance for COVID-19 in India through detection of genetic material of SARS-CoV-2. *Science of The Total Environment*, **746**, 141326, https://doi.org/10.1016/j.scitotenv.2020.141326

Lahrich S., Laghrib F., Farahi A., Bakasse M., Saqrane S. and El Mhammedi M. A. (2021). Review on the contamination of wastewater by COVID-19 virus: impact and treatment. *Science of The Total Environment*, **751**, 142325, https://doi.org/10.1016/j.scitotenv.2020.142325

La Rosa G., Bonadonna L., Lucentini L., Kenmoe S. and Suffredini E. (2020). Coronavirus in water environments: occurrence, persistence and concentration methods-A scoping review. *Water Research*, **179**, 115899, https://doi.org/10.1016/j.watres.2020.115899

Larsen D. A. and Wigginton K. R. (2020). Tracking COVID-19 with wastewater. *Nature Biotechnology*, **38**(10), 1151–1153, https://doi.org/10.1038/s41587-020-0690-1

Li Q., Guan X., Wu P., Wang X., Zhou L., Tong Y., Ren R., Leung K. S. M., Lau E. H. Y., Wong J. Y., Xing X., Xiang N., Wu Y., Li C., Chen Q., Li D., Liu T., Zhao J., Liu M., Tu W., Chen C., Jin L., Yang R., Wang Q., Zhou S., Wang R., Liu H., Luo Y., Liu Y., Shao G., Li H., Tao Z., Yang Y., Deng Z., Liu B., Ma Z., Zhang Y., Shi G., Lam T. T. Y., Wu J. T., Gao G. F., Cowling B. J., Yang B., Leung G. M. and Feng Z. (2020). Early transmission dynamics in wuhan, China, of novel coronavirus–infected pneumonia. *New England Journal of Medicine*, **382**, 1199–1207. https://doi.org/10.1056/NEJMoa2001316

Liu Y., Gayle A. A., Wilder-Smith A. and Rocklöv J. (2020). The reproductive number of COVID-19 is higher compared to SARS coronavirus. *Journal of Travel Medicine*, 1–4. https://doi.org/10.1093/jtm/taaa021

Long, Q.X., Tang, X.J., Shi, Q.L., Li, Q., Deng, H.J., Yuan, J., Hu, J.L., Xu, W., Zhang, Y., Lv, F.J., Su, K., Zhang, F., Gong, J., Wu, B., Liu, X.M., Li, J.J., Qiu, J.F., Chen, J., and Huang, A.L. (2020). Clinical and immunological assessment of asymptomatic SARS-CoV-2 infections. *Nature Medicine*, **26**(8), 1200–1204.

Lu D., Huang Z., Luo J., Zhang X. and Sha S. (2020). Primary concentration–The critical step in implementing the wastewater based epidemiology for the COVID-19 pandemic: A mini-review. *Science of The Total Environment*, **747**, 141245, https://doi.org/10.1016/j.scitotenv.2020.141245

Majumdar S. A., Nayak A. K. and Misra S. N. (2019). The perception study of water supply and drainage in the city of Bhubaneswar. *International Journal of Civil Engineering and Technology*, **10**, 664–670.

McDermott C. V., Alicic R. Z., Harden N., Cox E. J. and Scanlan J. M. (2020). Put a lid on it: are faecal bio-aerosols a route of transmission for SARS-CoV-2? *Journal of Hospital Infection*, **105**(3), 397–398, https://doi.org/10.1016/j.jhin.2020.04.024

Medema G., Heijnen L., Elsinga G., Italiaander R. and Brouwer A. (2020). Presence of SARS-Coronavirus-2 in sewage and correlation with reported COVID-19 prevalence in the early stage of the epidemic in the Netherlands. *Environmental Science & Technology Letters*, **7**(7), 511–516.

Mohapatra, S., Menon, N.G., Mohapatra, G., Pisharody, L., Pattnaik, A., Menon, N.G., Bhukya, P.L., Srivastava, M., Singh, M., Barman, M.K., Gin, K.Y.H. and Mukherji S. (2020). The novel SARS-CoV-2 pandemic: possible environmental transmission, detection, persistence and fate during wastewater and water treatment. *Science of the Total Environment*, **765**, 142746.

Morera S., Corominas L., Rigola M., Poch M. and Comas J. (2017). Using a detailed inventory of a large wastewater treatment plant to estimate the relative importance of construction to the overall environmental impacts. *Water Research*, **122**, 614–623, https://doi.org/10.1016/j.watres.2017.05.069

Morvan M., Jacomo A. L., Souque C., Wade M., Hoffmann T., Pouwels K. K., Singer A., Bunce J., Engeli A., Grimsley J., O'Reilly K. and Danon L. (2021). Estimating SARS-CoV-2 prevalence from large-scale wastewater surveillance: insights from combined analysis of 44 sites in England.

Nemudryi, A., Nemudraia, A., Wiegand, T., Surya, K., Buyukyoruk, M., Cicha, C., Vanderwood, K.K., Wilkinson, R. and Wiedenheft, B. (2020). Temporal detection and phylogenetic assessment of SARS-CoV-2 in municipal wastewater. *Cell Reports Medicine*, **1**(6), 100098. https://doi.org/10.1016/j.xcrm.2020.100098

Office for National Statistics (ONS) (2021). Coronavirus (COVID-19) Infection Survey, UK: 23 July 2021 – latest release. Available from: https://www.ons.gov.uk/peoplepopulationandcommunity/healthandsocialcare/conditionsanddiseases/bulletins/coronaviruscovid19infectionsurveypilot/23july2021 (Accessed 25th July 2021).

Oller I., Malato S. and Sánchez-Pérez J. (2011). Combination of advanced oxidation processes and biological treatments for wastewater decontamination–a review. *Science of the Total Environment*, **409**(20), 4141–4166, https://doi.org/10.1016/j.scitotenv.2010.08.061

Ong S. W. X., Tan Y. K., Chia P. Y., Lee T. H., Ng O. T., Wong M. S. Y. and Marimuthu K. (2020). Air, surface environmental, and personal protective equipment contamination by severe acute respiratory syndrome coronavirus 2 (SARS-CoV-2) from a symptomatic patient. *JAMA*, **323**(16), 1610–1612. https://doi.org/10.1001/jama.2020.3227

Peccia J., Zulli A., Brackney D. E., Grubaugh N. D., Kaplan E. H., Casanovas-Massana A., Ko A. I., Malik A. A., Wang D., Wang M. and Omer S. B. (2020). Measurement of SARS-CoV-2 RNA in wastewater tracks community infection dynamics. *Nature Biotechnology*, **38**(10), 1164–1167, https://doi.org/10.1038/s41587-020-0684-z

Polo D., Quintela-Baluja M., Corbishley A., Jones D. L., Singer A. C., Graham D. W. and Romalde J. L. (2020). Making waves: wastewater-based epidemiology for COVID-19–approaches and challenges for surveillance and prediction. *Water Research*, **186**, 116404, https://doi.org/10.1016/j.watres.2020.116404

Ramírez-Castillo F. Y., Loera-Muro A., Jacques M., Garneau P., Avelar-González F. J., Harel J. and Guerrero-Barrera A. L. (2015). Waterborne pathogens: detection methods and challenges. *Pathogens*, **4**(2), 307–334.

Randazzo W., Truchado P., Cuevas-Ferrando E., Simón P., Allende A. and Sánchez G. (2020). SARS-CoV-2 RNA in wastewater anticipated COVID-19 occurrence in a low prevalence area. *Water Research*, **181**, 115942, https://doi.org/10.1016/j.watres.2020.115942

Saba B., Hasan S. W., Kjellerup B. V. and Christy A. D. (2021). Capacity of existing wastewater treatment plants to treat SARS-CoV-2. A review. *Bioresource Technology Reports*, **15**, 100737, https://doi.org/10.1016/j.biteb.2021.100737

Szulc P., Kasprzak J., Dymaczewski Z. and Kurczewski P. (2021). Life cycle assessment of municipal wastewater treatment processes regarding energy production from the sludge line. *Energies*, **14**(2), 356, https://doi.org/10.3390/en14020356

Teymoorian T., Teymourian T., Kowsari E. and Ramakrishna S. (2021). Direct and indirect effects of SARS-CoV-2 on wastewater treatment. *Journal of Water Process Engineering*, **42**, 102193, https://doi.org/10.1016/j.jwpe.2021.102193

Valizadeh S., Younesi H. and Bahramifar N. (2016). Highly mesoporous K2CO3 and KOH/activated carbon for SDBS removal from water samples: batch and fixed-bed column adsorption process. *Environmental Nanotechnology, Monitoring & Management*, **6**, 1–13.

Vambol, S., Vambol, V., Mozaffari, N., Mahmoud, A.E.D., Ramsawak, N., Mozaffari, N., Ziarati, P. and Khan, N.A. (2021). Comprehensive insights into sources of pharmaceutical wastewater in the biotic systems. *Pharmaceutical Wastewater Treatment Technologies*, 17–46.

Venugopal A., Ganesan H., Raja S. S. S., Govindasamy V., Arunachalam A., Narayanasamy A., Sivaprakash P., Rahman, P.K., Gopalakrishnan, A.V., Siama, Z. and Vellingiri, B. (2020). Novel wastewater surveillance strategy for early detection of coronavirus disease 2019 hotspots. *Current Opinion in Environmental Science & Health*, **17**, 8–13.

Novel wastewater surveillance strategy for early detection of coronavirus disease 2019 hotspots. Current Opinion in Environmental Science & Health, 17, 8-13.

Wang C., Horby P. W., Hayden F. G. and Gao G. F. (2020a). A novel coronavirus outbreak of global health concern. *The Lancet*, **395**(10223), 470–473, https://doi.org/10.1016/S0140-6736(20)30185-9

Wang X. W., Li J. S., Jin M., Zhen B., Kong Q. X., Song N., Xiao W. J., Yin J., Wei W., Wang G. J., Si B. Y., Guo B. Z., Liu C., Ou G. R., Wang M. N., Fang T. Y., Chao F. H. and Li J. W. (2005b). Study on the resistance of severe acute respiratory syndrome-associated coronavirus. *Journal of Virological Methods*, **126**, 171–177, https://doi.org/10.1016/j.jviromet.2005.02.005

Wang, D., Hu, B., Hu, C., Zhu, F., Liu, X., Zhang, J., Wang, B., Xiang, H., Cheng, Z., Xiong, Y., Zhao, Y., Li, Y., Wang, X., Peng, Z. (2020c). Clinical characteristics of 138 hospitalized patients with 2019 novel coronavirus–infected pneumonia in Wuhan, China. *JAMA*, **323**(11), 1061–. https://doi.org/10.1001/jama.2020.1585

Wang J., Feng H., Zhang S., Ni Z., Ni L., Chen Y., Zhuo L., Zhong Z. and Qu T. (2020d). SARS-CoV-2 RNA detection of hospital isolation wards hygiene monitoring during the Coronavirus Disease 2019 outbreak in a Chinese hospital. *International Journal of Infectious Diseases*, **94**, 103–106, https://doi.org/10.1016/j.ijid.2020.04.024

Wang, J., Shen, J., Ye, D., Yan, X., Zhang, Y., Yang, W., Li, X., Wang, J., Zhang, L. and Pan, L. (2020e). Disinfection technology of hospital wastes and wastewater: suggestions for disinfection strategy during coronavirus disease 2019 (COVID-19) pandemic in China. *Environmental Pollution*, **262**, 114665, https://doi.org/10.1016/j.envpol.2020.114665

WHO (2020a). Status of Environmental Surveillance for SARS-CoV-2 Virus.

WHO (2020b). Summary Table for Available Protocols.

WHO (2020c). Water, Sanitation, Hygiene and Waste Management for the COVID-19 Virus.

Wigginton K. R., Ye Y. and Ellenberg R. M. (2015). Emerging investigators series: the source and fate of pandemic viruses in the urban water cycle. *Environmental Science: Water Research & Technology*, **1**(6), 735–746, https://doi.org/10.1039/C5EW00125K

Wilkinson A., Hill M. and Gollan P. (2001). The sustainability debate. *International Journal of Operations & Production Management*, **21**(12), 1492–1502.

Wong M. V. M., Hashsham S. A., Gulari E., Rouillard J.-M., Aw T. G. and Rose J. B. (2013). Detection and characterization of human pathogenic viruses circulating in community wastewater using multi target microarrays and polymerase chain reaction. *Journal of Water and Health*, **11**(4), 659–670. https://doi.org/10.2166/wh.2013.322

Wu, F., Xiao, A., Zhang, J., Moniz, K., Endo, N., Armas, F., Bonneau, R., Brown, M.A., Bushman, M., Chai, P.R., Duvallet, C., Erickson, T.B., Foppe, K., Ghaeli, N., Gu, X., Hanage, W.P., H Huang, K.H., Lee, W.L., Matus, M., McElroy, K.A., Nagler, J., Rhode, S.F., Santillana, M., Tucker, J.A., Wuertz, S., Zhao, S., Thompson, J., and Alm, E.J. (2020a). SARS-CoV-2 titers in wastewater foreshadow dynamics and clinical presentation of new COVID-19 cases. *medRxiv : The Preprint Server for Health Sciences*, https://doi.org/10.1101/2020.06.15.20117747

Wu F., Zhang J., Xiao A., Gu X., Lee W. L., Armas F., Kauffman K., Hanage W., Matus M., Ghaeli N. and Alm E. J. (2020b). SARS-CoV-2 titers in wastewater are higher than expected from clinically confirmed cases. *Msystems*, **5**(4), e00614–20.

Wu, Y., Guo, C., Tang, L., Hong, Z., Zhou, J., Dong, X., Yin, H., Xiao, Q., Tang, Y., Qu, X., Kuang, L., Fang, X., Mishra, N., Lu, J., Shan, H., Jinag, G., and Huang, X. (2020c). Prolonged presence of SARS-CoV-2 viral RNA in faecal samples. *The Lancet Gastroenterology & Hepatology*, **5**(5), 434–435, https://doi.org/10.1016/S2468-1253(20)30083-2

Wurtzer S., Marechal V., Mouchel J. M. and Moulin L. G. (2020). Time course quantitative detection of SARS-CoV-2 in Parisian wastewaters correlates with COVID-19 confirmed cases medRxiv. https://www.medrxiv.org/content/10.1101/2020.04.12.20062679v1 (accessed 28 October 2021)

Xiao F., Tang M., Zheng X., Liu Y., Li X. and Shan H. (2020). Evidence for gastrointestinal infection of SARS-CoV-2. *Gastroenterology*, **158**(6), 1831–1833.e3. https://doi.org/10.1053/j.gastro.2020.02.055

Yeo C., Kaushal S. and Yeo D. (2020). Enteric involvement of coronaviruses: is faecal–oral transmission of SARS-CoV-2 possible? *The Lancet Gastroenterology & Hepatology*, **5**(4), 335–337, https://doi.org/10.1016/S2468-1253(20)30048-0

Young P. N., Buchanan N. and Fallowfield H. J. (2016). Inactivation of indicator organisms in wastewater treated by a high rate algal pond system. *Journal of Applied Microbiology*, **121**(2), 577–586.

Yu D., Zhai J., Liu C., Zhang X., Bai L., Wang Y. and Dong S. (2017). Small microbial three-electrode cell based biosensor for online detection of acute water toxicity. *ACS Sensors*, **2**(11), 1637–1643.

Zaneti R. N., Girardi V., Spilki F. R., Mena K., Westphalen A. P. C., da Costa Colares E. R., Pozzebon A. G. and Etchepare R. G. (2021). Quantitative microbial risk assessment of SARS-CoV-2 for workers in wastewater treatment plants. *Science of The Total Environment*, **754**, 142163, https://doi.org/10.1016/j.scitotenv.2020.142163

Zhang N., Gong Y., Meng F., Bi Y., Yang P. and Wang F. (2020a). Virus shedding patterns in nasopharyngeal and fecal specimens of COVID-19 patients. *MedRxiv*.

Zhang N., Gong Y., Meng F., Shi Y., Wang J., Mao P., Chuai X., Bi Y., Yang P. and Wang F. (2020b). Comparative study on virus shedding patterns in nasopharyngeal and fecal specimens of COVID-19 patients. *Science China Life Sciences*, 10–15.

Zhang D., Ling H., Huang X., Li J., Li W., Yi C., Zhang T., Jiang Y., He Y., Deng S. and Qu J. (2020c). Potential spreading risks and disinfection challenges of medical wastewater by the presence of Severe Acute Respiratory Syndrome Coronavirus 2 (SARS-CoV-2) viral RNA in septic tanks of Fangcang Hospital. *Science of the Total Environment*, **741**, 140445, https://doi.org/10.1016/j.scitotenv.2020.140445

Zhang X., Lardizabal A., Silverman A. I., Vione D., Kohn T., Nguyen T. H. and Guest J. S. (2020d). Global sensitivity analysis of environmental, water quality, photoreactivity, and engineering design parameters in sunlight inactivation of viruses. *Environmental Science & Technology*, **54**(13), 8401–8410, https://doi.org/10.1021/acs.est.0c01214

Zhang, Y., Chen, C., Zhu, S., Shu, C., Wang, D., Song, J., Song, Y., Zhen, W., Feng, Z., Wu, G., Xu, J., and Xu, W. (2020e). Isolation of 2019-nCoV from a stool specimen of a laboratory-confirmed case of the coronavirus disease 2019 (COVID-19). *China CDC weekly*, **2**(8), 123–124, https://doi.org/10.46234/ccdcw2020.033

Zhao L., Dai T., Qiao Z., Sun P., Hao J. and Yang Y. (2020). Application of artificial intelligence to wastewater treatment: a bibliometric analysis and systematic review of technology, economy, management, and wastewater reuse. *Process Safety and Environmental Protection*, **133**, 169–182.

Zhu N., Zhang D., Wang W., Li X., Yang B., Song J., Zhao X., Huang B., Shi W., Lu R. and Niu P. (2020). A novel coronavirus from patients with pneumonia in China, 2019. *New England Journal of Medicine*, **382**(8), 727–733. https://doi.org/10.1056/NEJMoa2001017

doi: 10.2166/9781789062625_0053

Chapter 4

Bioremediation of pharmaceutical effluent by food industry and agricultural waste biomass

Parisa Ziarati[1], Maryam Mokhtarzadeh[2], Viola Vambol[3,4], Sergij Vambol[5], Barbara Sawicka[6], Alaa El Din Mahmoud[7,8] and Nadeem A. Khan[9,10]*

[1]*Medicinal Chemistry Department, Tehran Medical Sciences, Islamic Azad University, Tehran, Iran*
[2]*Department of Applied Chemistry, Faculty of Pharmaceutical Chemistry, Tehran Medical Sciences, Islamic Azad University, Tehran, Iran*
[3]*Department of Environmental Engineering and Geodesy, University of Life Sciences in Lublin, Lublin, Poland*
[4]*Department of Applied Ecology and Nature Management, National University 'Yuri Kondratyuk Poltava Polytechnic', Poltava, Ukraine*
[5]*Life Safety Department, State Biotechnological University, Kharkiv, Ukraine*
[6]*Department of Specific Plant Cultivation, University of Life Sciences in Lublin, Lublin, Poland*
[7]*Environmental Sciences Department, Faculty of Science, Alexandria University, Alexandria 21511, Egypt*
[8]*Green Technology Group, Faculty of Science, Alexandria University, Alexandria 21511, Egypt*
[9]*Department of Civil Engineering, Jamia Millia Islamia, New Delhi 110025, India*
[10]*Department of Civil Engineering, Mewat Engineering College, Nuh, India-122107*
**Corresponding author. E-mail: violavambol@gmail.com*

ABSTRACT

The sustainable removal of heavy metals from pharmaceutical effluent is still a major challenge to scientists and engineers despite there being numerous treatment technologies. Herein, the bio-adsorption of cobalt and nickel metal ions have been studied using cheap and readily available tomato pomace waste (TPW) to investigate various controlling parameters. These include particle size, doses, initial concentration of the studied heavy metals, contact time and pH. TPW as a residue can spoil quickly and is considered an environmental nuisance generated from the paste of manufacturing plants (tomato paste factories), due to having many functional groups for adsorbing heavy metals. The 920 samples studied were analyzed for two weeks and the heavy metal contents, nickel and cobalt, in each studied sample were analyzed by the wet digestion method and determined by inductively coupled plasma mass spectrometry (ICP-MS). Data analysis of variance revealed that the concentration of cobalt and nickel decreased to 63.93% of the original for nickel and 69.98% of the original for cobalt after 6 h in the presence of tomato biomass adsorbent which can easily be removed from pharmaceutical-contaminated effluent.

Keywords: adsorption, detoxification, heavy metals, pharmaceutical effluent, tomato pomace

CONTENTS

4.1 INTRODUCTION

Water pollution is divided into three categories: industrial, urban and agricultural pollution. Urban water pollution actually includes effluents from homes and businesses. Separately, it is worth paying attention to the treatment of medical and pharmaceutical wastewater (Khan *et al.*, 2019a, 2019b, 2020a, 2020b, 2021). So far, the main purpose of research on such effluents has been to reduce particulate matter, hard materials, solubility of inorganic compounds, and so on (Mahmoud *et al.*, 2021a; Plazzotta *et al.*, 2017; Ziarati *et al.*, 2016).

Nowadays, water pollution by heavy metals has become a major concern for human health that developing and industrialized countries face. Heavy metals are a health problem due to the serious and dangerous complications and problems they have always posed for the environment and humans (El Din Mahmoud & Fawzy, 2016; Ziarati *et al.*, 2015). The accumulation of metals in animals and plants, in addition to the serious damage and adverse effects on the health of these organisms, has made the consumption of their products dangerous for the final consumer, namely humans (Jallilian & Ziarati, 2016; Mahmoud *et al.*, 2016; Shirkhan, *et al.*, 2022; Tajik *et al.*, 2020). These dangers include disorders of the nervous, renal and circulatory systems.

Heavy metals, as a hazardous problem of various dimensions, can seriously endanger human life and other living organisms and cause acute and chronic poisoning (Badr *et al.*, 2020). Heavy metals have the ability to accumulate, such that their amount does not decrease, and remain in the soil for thousands of years (Mehrarad *et al.*, 2016; Shokri *et al.*, 2016; Ziarati *et al.*, 2020a, 2020b).

After entering the body, heavy metals are not excreted from the body, but accumulate in tissues such as fat, bone, joints, and muscles, eventually causing problems such as neurological disorders, various types of cancer, abortion, damage to the liver, kidneys and brain, arthritis, hair loss, respiratory disorders and osteoporosis in the human body (Ziarati *et al.*, 2020c). Therefore, contamination by this product can endanger human health. Heavy metals such as nickel, lead and mercury are dangerous toxins in the environment around us. These metals enter the natural cycle through water, soil and air from various natural and artificial sources and cause dangerous effects. Therefore, they are considered as a serious danger to the survival of living beings (Arancon *et al.*, 2013; Garadaghli, 2021; Ziarati *et al.*, 2019a, 2019b, 2019c).

Currently, the removal of heavy metals from municipal, industrial and pharmaceutical wastewaters is one of the most important aspects of water treatment and it needs to be done as effectively as possible because heavy metals are nonbiodegradable and their bioaccumulations are very dangerous (Khan *et al.*, 2020a, 2020b). Contaminated effluents should be removed before discharge (Mahmoud, 2020a), as contact with these metals can have very unpleasant consequences, such as neurological disorders, cellular aging, liver and kidney failure, and so on (Alimardan *et al.*, 2016; Mahmoud *et al.*, 2020; Sabiiti, 2011).

Studies have shown that fruit wastes are very effective in absorbing heavy metals due to their cellulosic, carbon and silica compounds that result in a high ability to absorb metal ions. Agricultural wastes are both abundant and available and have no specific use (Fungaro *et al.*, 2021; Mahmoud, 2020b). Since the late 1930s, the use of natural biosorbents such as raw plant residues has evolved and the method of chemical modification (using fruit waste) has replaced the method of physical activation (activated carbon) with advantages such as higher adsorption capacity and equilibrium time (Fatahi *et al.*, 2020; Tajik *et al.*, 2020).

Using sorbents is one of the most promising methods to remove contaminants. Tomato pulp is a plant residue and in this study its adsorption capacity for selected contaminants in water has been investigated (Ziarati *et al.*, 2021). In order to achieve efficient, effective and especially cheap adsorbents, the path of studies of adsorption of heavy metals to natural materials (such as some waste products from industrial and agricultural operations) has changed, the natural materials used are known as biomass (Ziarati *et al.*, 2021).

The terms wood biomass, short cycle wood products, agricultural wastes, short cycle plant species, wood wastes, industrial wastes, wastepaper, sawdust, bio-soils, grass, food processing wastes, aquatic plants, are wastes containing algae and other groups of substances. In fact, biomass is a simple term for all organic matter of biological origin (Arabian *et al.*, 2020; Seifollahi *et al.*, 2017; Ziarati *et al.*, 2020d).

Biosorbents have the ability to remove metal from solutions and can also be used to reduce their concentration (Fatahi *et al.*, 2020; Shirkhan *et al.*, 2022; Tajik *et al.*, 2020). They can effectively separate ppb to ppm of heavy metal ions from the surface of dissolved metal ions from complex dilute solutions with high efficiency and speed. Therefore, biosorbents are an ideal candidate for the treatment of wastewater with a large volume and low concentration of heavy metals. Some of the advantages of biosorption over conventional treatment methods include low cost, high efficiency for solutions with low concentrations of heavy metals, minimal amount of chemical or biological sludge created, no need for nutrients, and the possibility of reactivation of the adsorbent and metal recovery (Mahmoud, 2020c; Mousavi *et al.*, 2014; Ziarati, 2012; Ziarati & Tosifi, 2014).

The idea of utilizing agro/food waste as natural and available adsorbents to remove heavy metals from contaminated soils and wastewater is not new. Many studies have been conducted in this approach in recent years (Jafari *et al.*, 2016; Mahmoud *et al.*, 2018, 2021b; Shahsavan-Davoudi & Ziarati, 2020; Zahirnejad *et al.*, 2017).

In 2020, Shahsavan and Ziarati studied the adsorption properties of grapefruit peels for removal the arsenic, cobalt, lead, cadmium and chromium from pharmaceutical effluent in the Faculty of Pharmaceutical Chemistry, Tehran Medical Sciences in Azad University. They found that by increasing the size of biosorbent particles, absorption decreased and by decreasing the initial pH, time of contact of the solution by adsorbents, and temperature, the percentage of toxic metals significantly increased. Biosorbents treated with 1% tartaric acid at maximum capacity removed 10.01 mg/L of cadmium from the effluent, quickly removing it from the aqueous solution. Thermodynamic studies showed that the adsorption of cadmium on grapefruit-biosorbent is a spontaneous and exothermic process and the results revealed that grapefruit peel can be considered as an effective adsorbent (Alimardan *et al.*, 2016; Shahsavan-Davoudi & Ziarati, 2020; Ziarati & Mirmaohammad Makki, 2015). Tavakoli-Hosseinabady *et al.* (2018) investigated removal yields of lead, nickel and cadmium contents of edible vegetables using apricot shell. Thus, it was found that detoxification of heavy metals in plants, a serious issue before being used for food processing and human consumption, is better achieved with the apricot treatment process. The potential for developing heavy metal related illnesses and chronic degenerative conditions from various plants, including parsley (*Petroselinum crispum*); holy basil (*Ocimum tenuiflorum*); *Coriandrum sativum*; and Cress (*Lepidium sativum*), can be avoided by using apricot shell treatment in the soil. The removal of heavy metals from the contaminated agricultural soil, in Yazd city, was achieved using apricot kernel hard shell. Different adsorption parameters, such as adsorbent dose, adsorbent particle size and plant growth time, were examined. Composite soil samples were collected from four agricultural areas growing leafy vegetables in Yazd city, under similar controlled physical conditions including pH, light, irrigation and humidity. The apricot kernel had the ability to significantly adsorb heavy metals from the contaminated soil, and leafy vegetables were cultivated by different contents of bio-adsorbent (Tavakoli-Hosseinabady *et al.*, 2018).

During 2018–2021, Ziarati *et al.* studied methods for the removal of toxic heavy metals by biosorbents. The main purpose of this article is to evaluate the potential, reliability and effectiveness of using agricultural waste to significantly eliminate or reduce heavy metal contamination in food and water

(Ziarati *et al.*, 2020a). In the current study, the main aim is to adsorb toxic and heavy metals using cheap and readily available tomato pomace waste (TPW), an environmental residue which can spoil quickly and is considered an environmental nuisance generated from the paste of manufacturing plants (tomato paste factories). The purpose of this adsorption is to reduce the hazardous heavy metals nickel and cobalt in the effluent of pharmaceutical laboratories, and prevent the TPW from providing a breeding place for a variety of pests such as flies and mosquitoes, which are hosts of disease-causing organisms.

Tomato with the scientific name of *Solanum lycopersicum* is one of the red vegetables or fruits (Caluya *et al.*, 2003; Heuzé *et al.*, 2015; Ziarati & Mirmaohammad Makki, 2014). This plant is native to South and Central America and was transferred to other parts of the world during the Spanish colonial period. Different types of this plant are grown all over the world today. Fresh tomato by-products have been considered an environmental nuisance for a long time (Del Valle *et al.*, 2006). In some countries, the waste is dumped in water bodies near the factory or left to accumulate on the site of production. The material spoils quickly, emits a very foul odor and turns into a place for the formation of environmental hazards (León-García *et al.*, 2017).

Due to its economic importance, this plant is the subject of many studies and is known in genetics as one of the model plants. Research on this plant in 1990 led to the production of the first transgenic species permitted for consumption and trade in the United States (Banerjee *et al.*, 2017).

Tomato pomace is one of the by-products of the tomato paste industry, which, depending on the processing method and characteristics of the raw tomatoes, includes different proportions of skin, seeds and small amounts of tomato meat. Tomato pulp is one of the by-products that is obtained during the production process of tomato sauce and puree (Ehsani *et al.*, 2021).

The perception is that tomato pomace, due to its many functional groups, probably can be a very valuable adsorbent that is able to reduce the hazardous heavy metals nickel and cobalt from the effluent of pharmaceutical laboratories.

The novelties of this study are:

(1) Utilizing green and natural bio-adsorbent in regaining the considerable potential of food waste, using it as a source of many green products, and opening up the possibility for their extraction and purification;
(2) Using new methods and assistance in solving the agricultural/food waste problems by government agencies can greatly help reduce heavy metals pollution and improve the production cycle.

4.2 PROSPECTIVE METHODS FOR HEAVY METALS REMOVAL IN WASTEWATER USING PLANT ADSORBENTS

Purification of natural and waste waters by removal of heavy metals is effectively achieved using various sorbents. The sources of raw materials for obtaining such materials are very diverse. Wastes from biomass processing constitute a special group. Active porous materials from plant waste are of significant practical interest for solving a number of environmental problems in industrially saturated regions: wastewater treatment, gas emissions, soil contamination, and so on. It is known that such sorbents are capable of extracting heavy metals from water waste, thereby reducing the anthropogenic load on natural water bodies (rivers, lakes, etc.).

Of particular interest is the problem of purifying pharmaceutical wastewaters by removing heavy metal ions (Khan *et al.*, 2019b). Wastes in the form of biochemical compounds in combination with heavy metal ions are quite hazardous substances (Husain Khan *et al.*, 2020). It follows from this that the development of new methods of purification for such pollutants and the search for new sorbents on a plant basis are urgent.

The use of sorbents from biomass waste is economically justified for small industries close to the sources of available raw materials. Analysis of the structure of the raw material base of plant wastes shows that the most promising wastes, in terms of their availability and scale for obtaining sorption materials, are agricultural and wood-processing waste.

4.2.1 Most common methods of wastewater treatment for removal of heavy metal ions

The current methods of wastewater treatment for removal of heavy metal ions (HMI) can be divided into reagent, biochemical, electrochemical, membrane, combined and sorption. The reagent method consists of obtaining poorly soluble substances with subsequent sedimentation. $Ca(OH)_2$, NaOH, Na_2S, ferrochrome slag, $FeSO_4$, and FeS_2 are used as reagents. For the deposition of metals, Ca $(OH)_2$ is widely used, precipitating metal ions according to the scheme: $Me^{n+} + nOH- = Me(OH)n$. Reagent purification of acid-alkaline waters is carried out using $Ca(OH)_2$ or Na_2CO_3 when the pH of the solution is brought to 9. Metal hydroxides are formed, as well as sorption of Men+ ions by the surface of the solid phase. The average hydraulic particle size is 0.2–0.4 mm. Since the product of solubility for hydroxocarbonates is lower, they are formed at lower pH values, but require more intensive consumption of soda. To separate the resulting suspensions, centrifugal or gravity separation, or flotation are used. For coagulation, iron or aluminum salts are used, as well as polyacrylamide – 0.1% of the solid phase content. The main disadvantages and problems of this method include the need to use a large number of reagents and the fact that the purification process is time-consuming and demanding. For its implementation, it is necessary to use a lot of equipment, and to create sludge storage facilities. An effective reagent for the removal of heavy metals is Na_2S, due to the extremely low solubility of heavy metal sulfides: $Me^{2+} + S_2^- = 2MeS$; $2Me^{3+} + 3S_2^- = 3Me_2S_3$. To accelerate the process of precipitation of metal sulfides, coagulants and flocculants are used. Colloidal particles of sulfides have a negative charge; therefore, $Al_2(SO_4)_2$ or $FeCl_3$, as well as their mixtures, are usually used as coagulants. Using mixtures of $Al_2(SO_4)_3$ and $FeCI_3$ in ratios from 1:1 to 1:2, a better coagulation result is achieved than when the reagents are used separately. For local wastewater treatment for removal of heavy metals (Cu, Ni, Pb, Mn, Co, Fe, Zn, Cr, etc.), a chelating agent with dithioamino groups can be used. The resulting precipitate is then removed by precipitation or filtration. It is also possible to release heavy and non-ferrous metals from the wastewater of galvanic production by the biochemical method of sulfate-reducing bacteria. However, this method has a low cleaning efficiency.

Ion-exchange wastewater treatment for removal of heavy metal ions is becoming more widespread. From an economic point of view, the most expedient ion-exchange treatment is not for general production runoff, but for wastewater generated in individual technological processes and operations containing the least metals possible. Ion-exchange regeneration methods allow not only the complete extraction of non-ferrous, heavy and rare metals from waste solutions, but also obtaining regeneration products in the form of pure metal salts, suitable for reuse in production in order to prepare anew and adjust working electrolytes. In addition, the purified water obtained after ion-exchange treatment can in most cases be used as circulating water without additional treatment. When ion exchange is reversible, the processes of sorption and desorption are interrelated. The conditions affecting the sorption process also affect the desorption process, that is, they determine the ease and completeness of desorption of ions from the cation exchanger with a solution of one or another electrolyte, and the consumption and degree of use of the regenerating solution. The rate and completeness of leaching is determined not only by the properties of the sorbed ion (charge, size, degree of hydration, the ability to form complex ions), but also by the properties of the salt or acid ion of the eluent, its ion-exchange affinity for the ion exchanger of this type, and its concentrations.

Electrochemical methods include electrocoagulation (galvanic coagulation), electroflotation, electrolysis, and so on. When using electrochemical methods of purification by removal of metal ions, it is possible to achieve a residual concentration of metal ions below the maximum permissible concentration (MPC). Disadvantages of the method include high power consumption, loss of heavy metals with sludge, impossibility of reusing water, the need for preliminary dilution of effluents, passivation of anodes, consumption of soluble anodes, and the need for space for sludge storage. The advantages of the method include cleaning to the MPC level, low consumption of reagents, ease of operation, compactness of equipment, absence of secondary pollution, removal of fats and oils, and recycling of heavy metals.

The **electroflotation method** provides for the purification of electroplating wastewater by removal of ions of heavy metals to the maximum permissible concentration, and also removal of fats and oils. The disadvantages of the method are a small decrease in the salinity of the effluents, a scarce anode material, and a large consumption of energy resources. A method of electrodialysis is also proposed, which consists of the selective transfer of ions through partitions made of ion exchangers (membranes) under the action of an electric current. There are also methods of purification which combine the use of membranes and ion-exchange. Ion-exchange recovery of metals from wastewater allows a high degree of recovery of valuable substances. Synthetic ion-exchange resins have acquired the greatest practical importance for wastewater treatment. However, these methods are expensive and generate toxic secondary waste.

The prospect of using **plant adsorbent as a filtering** and purifying element is very promising. Of particular interest are plant substances that are a by-product or waste of food processing processes. It is the economic factor (the cost of the adsorbing substance) that has forced many researchers to pay attention to the removal of heavy metals from wastewater. For this method, research has mainly been focused on the use of combined sorbents based on agricultural processing wastes, sorbents of the woodworking industry and plant sorbents.

4.2.2 Most common methods of wastewater treatment for removal of cobalt and nickel ions

Nickel and cobalt metal salts are contained in wastewater from chemical-pharmaceutical, chemical, paint and varnish, textile and other industries. At the same time, whole branches of the technogenic sphere deserve special attention, where mining and processing factories, metallurgical, machine-building and metal-working enterprises operate. In these cases, the volume of wastewater discharge is several orders of magnitude higher than the previously mentioned industries.

The predominant treatment methods involve reagents for removing ions of the heavy metals nickel and cobalt from wastewater (BLOG, 2016). However, when acidic waters are treated with calcium oxide and sodium hydroxide, the ions of these heavy metals contained in the effluents are bound to form poorly soluble compounds. The composition of the salts depends on the pH of the medium. The most widely used reagent is calcium hydroxide. Precipitation of metals occurs in the form of hydroxides. The process is carried out at different pH values. The pH values corresponding to the onset of precipitation of hydroxides of various metals and complete precipitation depends on the nature of the metals, their concentration in the solution, temperature, and the content of impurities. For example, with the co-deposition of two or more metal ions at constant pH, better results are achieved than with the deposition of each of the metals separately. In this case, mixed crystals are formed and metal ions are adsorbed on the surface of the solid phase, due to which a more complete removal of some metals is achieved.

Purification of wastewater by removal of nickel is based on its separation from solution in the form of poorly soluble compounds. The precipitation of nickel hydroxide begins at pH = 6.7. The solubility product of $Ni(OH)_2$ is 2.0×10^{-18}. The concentration of nickel cations is extremely dependent on pH: at pH = 8.0, it is 23.5 mg/L and at pH = 9.5, only 0.5 mg/L. The solubility product of nickel carbonate is higher than that of hydroxide and equals 6.6×10^{-9}. The solubility of nickel hydroxide carbonate is practically zero, therefore, for the most complete removal of nickel, grade III lime, containing $CaCO_3$, should be used.

As noted above, the co-deposition of nickel and cobalt achieves better results than the deposition of each of the metals separately. The is due to the formation of mixed crystals and the adsorption of metal ions on the surface of the solid phase. To increase the degree of purification of wastewater containing heavy metals, it is proposed, after precipitation of their hydroxides with lime at a pH of 8.5 and separation of the sludge, to introduce a solution into the clarified wastewater in an amount 5–30 times higher than the stoichiometric norm. After separation of the sludge, wastewater with a low content of heavy metal ions can be returned to the water recycling system. The treatment of wastewater with alkaline reagents makes it possible to reduce the content of heavy metals in the solution to values comparable to the MPC for sanitary water bodies. However, when deeper cleaning is required, for example, for direct discharge into fishery reservoirs, cleaning with alkaline reagents does not give the required effect. Further removal of heavy metals is achieved when wastewater is treated with sodium sulfide. This is due to the fact that the solubility of heavy metal sulfides is much lower than the solubility of any other sparingly soluble compounds, such as hydroxides and carbonates.

In the studies carried out by Zakhvatkin (2012), the regularities of the deposition of cobalt and nickel ions from dilute solutions corresponding to the content of these metals in the waste streams of mining and metallurgical plants are considered. Commercial-scale sulfhydryl compounds – dithiocarbamates, ethyl xanthate, dithiophosphate, and dithiocarbamic acid derivatives – were tested as precipitants. The deposition was carried out at pH values of the medium that prevent the formation of cobalt and nickel hydroxides. It was shown that almost complete deposition of metals can be achieved using both ethyl xanthate and sodium ethyldithiocarbamate. The hydrophobic sludge formed in this way can be recovered by flotation in machines with low energy dissipation.

When considering pharmaceutical wastewater treatment technologies, pyrite may be used to remove small amounts of heavy metal ions. The process can be carried out by filtration of wastewater through granular pyrite or by direct injection of the powder into the wastewater. In addition to pyrite, sulfide of any other non-toxic metal can be used for this purpose, providing its solubility product is greater than solubility product of the sulfide of the metal extracted from the wastewater. Since hydroxides and sulfides of heavy metals form stable colloidal systems, coagulants and flocculants must be introduced to intensify the process of their precipitation into wastewater. Colloidal particles of sulfides have a negative charge; therefore, electrolytes with multiple charged cations are used for the coagulation of sulfides – usually aluminum or ferric sulfates. When hydroxides are coagulated, an electrolyte with multiple charged anions is required as the colloidal particles are positively charged. Sulfate ions are good hydroxide coagulants. In addition to electrolyte coagulation, in practice, mutual coagulation of colloidal solutions with opposite particle charges is often used. To accelerate the coagulation process, flocculants are used, mainly polyacrylamide. Its addition at a level of 0.01% of the dry matter weight increases the rate of precipitation of metal hydroxides by 2–3 times.

The disadvantage of such cleaning is the formation of a large amount of sludge which is difficult to drain. In addition, purified water contains a large amount of calcium salts, so it is difficult to use it in a recycled water supply. Based on this, it was proposed to treat the drain with calcium chloride and soda, sequentially, after settling it. In this case, the coprecipitation of metal carbonates with calcium carbonate occurs. The resulting crystalline precipitates of metal carbonates have an insignificant volume and are easily dewatered. At the same time, water softening occurs, which makes it possible to use it in the circulating (recycled) water supply system.

4.2.3 Purification of wastewater by removal of cobalt and nickel ions using acids

In the research conducted by Pimneva (2013), the processes of desorption of metal ions from the phosphoric acid cation exchanger KFP-12 were considered and solutions of acids HCL, H_2SO_4, NH_4Cl, and $(NH_4)_2SO_4$ were used. The process for washing out metal cations depends on the type of electrolyte, its ability to enter into complexation reactions with sorbed cations, the pH of the medium, and the concentration of metal ions in the eluate.

In aqueous solutions, hydrochloric and sulfuric acids behave like strong electrolytes. Sulfuric acid is a dibasic acid, the stepwise dissociation constants at a temperature of 25°C ($\mu = 0$) are equal to $K_1 = 1 \times 10^3$ and $K_2 = 1.2 \times 10^{-2}$. Depending on the concentration of sulfuric acid in solutions, the ionic composition changes. Sulfuric acid in solutions is less dissociated in solutions than hydrochloric acid. As a result of the different degrees of dissociation of these compounds, a difference in the concentration of cations and the pH of the medium is observed. For example, in the series of ammonium salts, the degree of dissociation and, therefore, the concentration of free ammonium ions decreases significantly.

The noted differences are significant and should have a great influence on the behavior of metal cations when they are washed out from the KFP-12 cation exchanger. For cations associated with functional groups of the cation exchanger by ionic bond, the concentration of the displacing ion plays a decisive role in the eluting solution. The weakly acidic cation exchanger KFP-12 has a high affinity for hydrogen ions and low for ammonium ions. This causes the difference in the leaching ability of solutions of mineral acids and ammonium salts in relation to metal ions, if the latter do not form chloride or sulfate complexes. Otherwise, the strength of the complex compounds will affect the leaching out. For example, copper (II), cobalt and nickel form anionic chloride complexes at hydrochloric acid concentrations of, respectively: 4.0, 6.5 and 10–11 mol/dm^3. The behavior of the cations bound to the ionogenic group by ion-coordination bond will be determined by the pH of the medium and the strength of the complexes formed by the interaction of the metal cation with the elution solution. In an acidic environment, the destruction of the metal complex with an ionic group is possible. The activity coefficients of hydrochloric and sulfuric acids and their ammonium salts in aqueous solutions at 25°C are shown in Table 4.1.

Table 4.1 The activity coefficients of hydrochloric and sulfuric acids and their ammonium salts in aqueous solutions at 25°C.

Electrolyte	Activity Coefficient at Electrolyte Concentration, mol/100 g					
	0.01	0.05	0.1	0.3	0.5	1.0
H_2SO_4	0.544	0.340	0.266	0.183	0.156	0.132
HCl	0.905	0.831	0.796	0.777	0.789	0.871
$(NH_4)_2SO_4$	–	–	0.770	0.687	0.649	0.603
NH_4Cl	–	–	0.423	0.300	0.248	0.183

In this case, the metal cations will be washed out by acid solutions, but the washout by solutions of neutral salts will not be observed. It should be added that solutions of mineral acids cannot always be effective desorbents. This is the case when the strength of complex compounds of a metal with an ionic group is very high even in acidic solutions. In these cases, the desorbing solution must contain an ion that can interact with the sorbed element to form a more stable complex compound as compared to the polymer complex. For example, indium is desorbed only by hydrochloric acid solutions. The effectiveness of hydrochloric acid is associated with the formation of indium anionic chloride complexes.

The considered properties of hydrochloric and sulfuric acids and their salts as eluents, taking into account the complexing properties of the cation exchanger and the ability of sorbed cations of elements to form neutral or anionic complexes, show that the difference in the behavior of metal cations at the stage of desorption should manifest itself when using solutions of ammonium chloride and sulfate. The ability of ionic groups to protonate should equalize the behavior of cations that do not form chloride or sulfate complexes during the desorption of the cation exchanger with solutions of hydrochloric and sulfuric acids. Everything noted is well supported by experimental data.

4.2.4 Isolation of nickel cations using colloidal sorbents

In Ganebnykh (2010), a method for purifying wastewater by removal of heavy metal compounds was proposed using the example of the process of separating nickel cations using colloidal sorbents. Natural and modified montmorillonites were used as sorbents, for which the adsorption isotherms of nickel ions were obtained, and the adsorption equilibrium constants and the capacity of the adsorption layer were calculated. The adsorption of nickel ions was carried out on modified montmorillonite hydrosols. The possibility of removing metal compounds to a low residual concentration regardless of their stability, the possibility of automating the process, the convenience of monitoring the purification process, and the absence of secondary pollution of the purified water are shown. Non-carbon clay sorbents (vermiculite, kaolinite, diatomite, montmorillonite, and others) were isolated from a wide variety of sorbents. They have a relatively high sorption capacity and selectivity.

When conducting research, in order to increase the sorption capacity, sorbents of natural origin were treated with modifiers for the formation of poorly soluble or complex compounds with metal cations. The concentration of the dispersed phase in the modified and unmodified hydrosols was 50 g/dm^3. As the object of the study, we used aqueous solutions of nickel (II) sulfate with a concentration of 2 to 40 mg/dm^3 by cation in the pH range of 7.0–9.5 model solutions.

The type of sorption isotherm depends on the pH of the solution, since along with sorption processes, the process of heterocoagulation of the formed particles of metal hydroxides with a highly developed surface of montmorillonite takes place. The concentration of $Ni(OH)_2$ hydroxide increases with an increase in pH and an increase in the initial concentration of metal in the solution. At

Table 4.2 Results of wastewater treatment of OJSC 'uralelectromed'.

Processing Method	Reagent	pH	Ni, mg/dm³
Source water	–	8.0–10.0	0.1–0.4
Cleaning according to the current technology	lime	9.8 ± 0.1	0.22 ± 0.08
Filtration followed by centrifugation	6.27	8.2 ± 0.1	<0.015
Contact clarification (quartz sand)	6.27	8.6 ± 0.1	<0.015

$pH \geq 8.5$, nickel hydroxide is completely absent, and all nickel is in ionic form (Ni^{2+} and $NiOH^{+}$). With an increase in pH, an increase in the value of the maximum adsorption is observed for all studied bentonite samples.

To obtain sorption isotherms, bentonite hydrosol (100 mg/dm³ in solid phase) was added to the model solution of nickel sulfate with vigorous stirring, and after 30 min with slow stirring, the solid phase was separated by centrifugation.

To describe the process of monomolecular adsorption of metal cations using colloidal sorbents in the studied pH range, the Langmuir equation was used:

$$A = A_{\infty} \frac{K \cdot C}{1 + K \cdot C} \tag{4.1}$$

where A is the value of adsorption, mg/g; A_{∞} is the capacity of the adsorption layer, mg/g; K is the constant of adsorption equilibrium, dm³/mg; and C is the equilibrium concentration in volume, mg/dm³.

According to the results of the study, it was concluded that the modification of montmorillonite can significantly increase the capacity of the adsorption layer. The maximum value of adsorption with respect to nickel at a concentration of the initial solution of 35.00 mg/L was shown by sample 6.27 (140 mg/g in terms of the weight of the solid sorbent). This sample was chosen for further studies of the sorption properties on industrial wastewaters of the Open Joint Stock Company (OJSC) 'Uralelectromed'. The experimental results show more than a 10-fold improvement in the degree of purification from nickel ions in comparison with the existing technology (Table 4.2).

The research results showed a high efficiency of using modified montmorillonites as sorbents for the extraction of nickel (II) ions from both model solutions and real wastewaters. Further studies of the sorption of these materials aim to obtain modified sorbents for use in existing technological schemes for wastewater treatment.

4.2.5 Purification of water by removal of cobalt ions using activated carbon fiber

Scientific research conducted by Valinurova *et al.* (2007) considered the sorption activity of activated carbon fiber (ACF) obtained from cellulose hydrate and the possibility of using carbon fiber and its modified samples for the sorption of cobalt ions from model mineralized aqueous solutions, as well as wastewater containing a significant amount of organic pollutants.

Researchers also note that sorption treatment methods are effective at low concentrations of harmful components, so it is advisable to use them for post-treatment of wastewater. Sorption on activated carbon is considered as an alternative effective method for removing heavy metal ions from solutions, especially at low metal concentrations. Therefore, the possibility of using sorption on activated carbon materials to remove cobalt ions is obvious, since their content in effluents is usually much less than that of other heavy metals (Rodionov *et al.*, 2000). The potential for using carbon fiber materials for the extraction of heavy metal ions is due to the combination of filtering and sorption properties, high specific surface area and developed porosity in one sorbent. To increase the selectivity of adsorption, various methods of modifying carbon fibers are widely used including by oxidation in the gas phase with ozone; liquid phase oxidation with nitric or sulfuric acid; thermal treatment in vapors of a modifying substance, for example, hydrogen sulfide; or by applying chelating reagents. These allow a wide range of selective chemisorbents to be obtained (Zverev, 1981).

In this work, we used an activated carbon fiber material made on the basis of cellulose hydrate, brand UVIS-AK, whose specific surface area is 1288.3 m^2/g. The limiting adsorption capacity for condensed benzene vapors is 0.59, for heptane it is 0.62, and for water it is 0 cm^3/g. The average pore diameter is 4–9 Å.

Sorption was carried out under static conditions with constant stirring for 2 days for initial solutions containing from 6 to 60 mg/L Co^{2+}. The volume of the investigated solutions was 25 mL. The weighed amount of activated carbon fiber was 0.02 g. The content of cobalt in solutions after sorption was determined photometrically using a nitroso-R-salt. Nitroso-R-salt is one of the most sensitive and selective reagents for the photometric determination of cobalt. With cobalt, the nitroso-R-salt forms a colored stable red complex with a metal to reagent ratio of 1:3.

The amount of sorbed cobalt was calculated from the difference in its content in solutions before and after sorption.

The degree of extraction of cobalt ions by the sorbent (K_u) was calculated by the formula (Mahmoud & Fawzy, 2015):

$$K_u = \left(C_o - C_p\right)/C_o \times 100\% \tag{4.2}$$

where, C_0 and C_p are the initial and equilibrium concentrations of cobalt ions in solution, mg/L.

The adsorption value (a) was calculated by the formula:

$$a = \left(C_o - C_p\right)/qV \tag{4.3}$$

where, q is the amount of the sorbent sample, g; and V is the solution volume, L.

Carbon fibers obtained on the basis of cellulose hydrate were used which determines their tendency to ion exchange with liquid media containing ions of various metals, including heavy ones. Sorbents with carboxyl functional groups on their surface are known to be weakly acidic cation exchangers (Zverev, 1981).

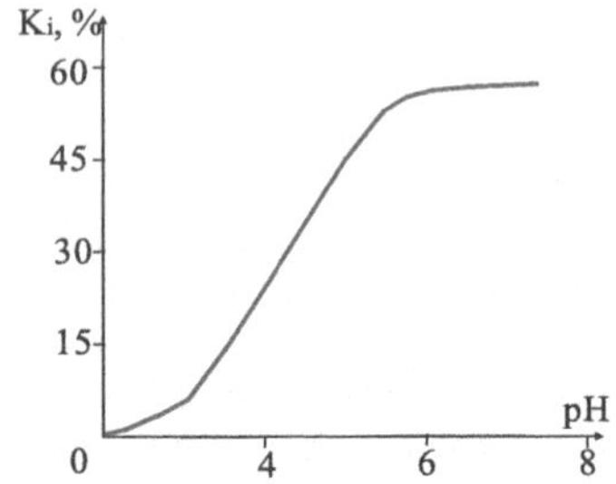

Figure 4.1 Extraction coefficient of cobalt ions by carbon fiber vs medium acidity. Figure created by the authors of this chapter based on the results of research by Zverev (1981).

One of the main factors determining the reactions of ion exchange between the sorbate and the sorbent containing weakly acidic and weakly basic groups is the pH value of the solution. Figure 4.1 shows the dependence of the degree of extraction on the acidity of the medium.

The change in pH in the test solutions was created by adding a small volume of solutions of hydrochloric acid and sodium hydroxide. An increase in the pH value contributed to an increase in the degree of extraction of cobalt ions from aqueous solutions. A change in pH from 1.8 to 10.6 increased the degree of extraction of cobalt ions by carbon fiber from 6 to 57%. An increase in pH promotes an increase in the degree of dissociation of acid sites of the sorbent and, as a consequence, a better adsorption of cobalt ions. The optimal neutral medium for the extraction of cobalt AHC was determined and selected.

The kinetics of the process is influenced not only by the chemical nature of the functional groups of the sorbent, but also by the concentration of ions in the solution. The kinetic dependences of the sorption of cobalt ions from solutions with the initial concentration of 24 and 12 mg/L are shown in Figure 4.2.

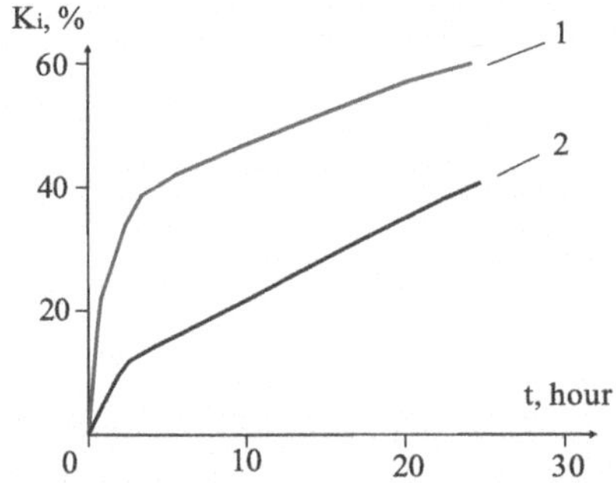

Figure 4.2 Kinetic dependences of cobalt ions sorption on carbon fiber; 1 – C(Co^{2+}) = 12 mg/L; 2 – C(Co^{2+}) = 24 mg/L. Figure created by the authors of this chapter based on the results of research by Zverev (1981).

Table 4.3 The influence of sodium ions on degree of cobalt extraction from AUV solution.

C_{Na^+}, mg/L	K_i, %	C_{Na^+}, mg/L	K_i, %
0	41.4	2760	21.4
180	22.3	3680	18.0
920	26.0	4600	17.4

It is obvious that with the higher initial concentration of cobalt ions in the solution, it takes longer establish sorption equilibrium, and the degree of extraction of metal ions decreases from 60 to 40%. The low rate of ion exchange on carboxyl-containing sorbents is explained by the low degree of their dissociation, especially if they are in the H-form.

The magnitude of the charge of the ion in the solution also affects the kinetics of exchange. The higher the ion charge, the lower the sorption rate due to steric factors. The values of the degree of extraction of cobalt ions by fiber, depending on the content of sodium ions in the solution are given in Table 4.3. With an increase in the concentration of sodium ions in the solution, the degree of extraction of cobalt decreases markedly. More mobile sodium ions diffuse faster to the acid sites of the sorbent than hydrated cobalt ions.

An increase in the sorption temperature leads to an acceleration of sorption processes. When the temperature rises by 10°C, the diffusion coefficient increases by 1.5–2.0 times. When the temperature rises by 15–20°C, the degree of extraction of cobalt ions by the fiber increases by 7%.

To increase the selectivity of sorption purification, the carbon fiber was modified by oxidizing with 70–73% nitric acid during boiling, and impregnating with solutions of nitroso-R-salt and sodium sulfide salts. Kinetic curves of CO^{2+} sorption on modified and initial unmodified fibers from solutions containing up to 600 mg/dm³ Na^+ are shown in Figure 4.3.

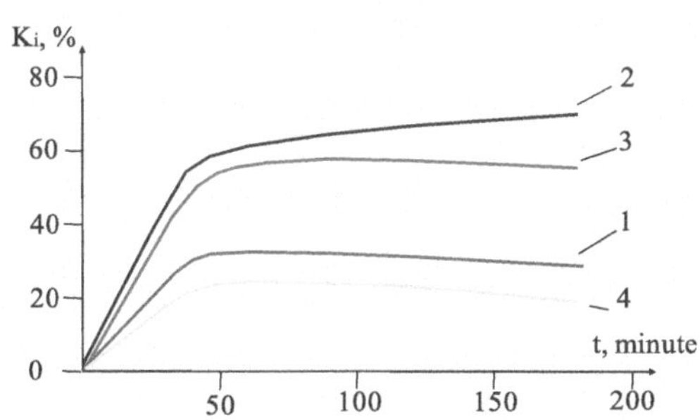

Figure 4.3 The Co^{2+} extraction coefficient from solutions with interfering Na^+ ions (600 mg/L), 1 – by virgin fiber and by fibers modified by: 2 – nitroso-R-salt; 3 – oxidation by nitric acid; 4 – solution of sodium sulfide. Figure created by the authors of this chapter based on the results of research by other authors.

It is obvious that the treatment of the fiber with a complexing salt significantly increases the efficiency and selectivity of sorption. The CO^{2+} extraction coefficient was increased to 71%. The traditional laboratory method for the production of fibrous oxidized coals is their oxidation with a solution of nitric acid. As can be seen from Figure 4.3, oxidation also promotes an increase in the sorption activity of carbon fiber with respect to cobalt ions, which is associated with an increase in the number of carboxyl groups on the sorbent surface, that is during liquid-phase oxidation with concentrated nitric acid, the cation-exchange properties of the fiber increased. The treatment with sodium sulfide did not lead to an improvement in the sorption properties of the sorbent; it is possible that during modification the acid groups on the fiber surface were blocked by sodium ions formed during salt hydrolysis. As a result, the coefficient of extraction of Co^{2+} + decreased by 7–10% in comparison with unmodified fiber (K_i=25–32%).

For the purification of industrial wastewater, active carbons of various grades are widely used, therefore, in this work, a comparison was made of the sorption activity of the carbon fiber under study with activated carbon of the BAU brand.

Figure 4.4 shows the isotherms of the sorption of cobalt ions on the initial carbon fiber, activated carbon of the BAU brand, and oxidized carbon fiber from aqueous solutions of cobalt sulfate at a pH of 7 and contact time of the solutions with the sorbent of 2 days.

It was found that carbon sorbents are characterized by significant sorption of cobalt ions. The isotherm of sorption of cobalt ions on AHC is located higher than on BAU activated carbon. The adsorption on the fiber is better, since the fibers are characterized by a higher permeability due to their small diameter. Oxidation of carbon fiber with concentrated nitric acid promotes an increase in the number of carboxyl groups on the sorbent surface, which is expressed in a significant increase in the adsorption of cobalt ions.

After purification of a sample of a catalytically oxidized sulfide-alkaline effluent containing oil products (25 mg/L), phenols (100 mg/L), sulfates (2300 mg/L), sulfides (5 mg/L) and Co^{2+} (20 mg/L), no cobalt was detected.

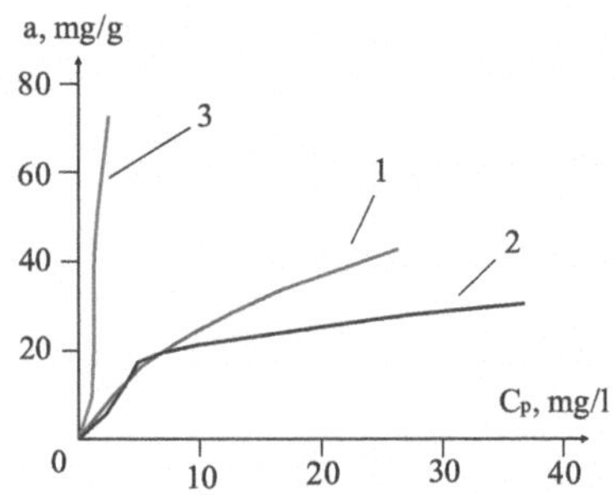

Figure 4.4 Isotherms of cobalt ions sorption on 1 – virgin carbon fiber, 2 – activated coal, and 3 – oxidized carbon fiber. Figure created by the authors of this chapter based on the results of research by other authors.

According to the research results, it was found that carbon cellulose hydrate fiber has a high sorption capacity for cobalt ions and can be used for removal of cobalt salts to achieve fine water purification. To increase the efficiency and selectivity of adsorption of cobalt ions from highly saline waters, it is recommended to use modified fibrous carbon materials.

4.3 MATERIALS AND RESEARCH METHODS

This study used a standard method for assessing the content of heavy metals. Previously, this approach was used to determine the adsorption properties of other plant biosorbents. The selection and preparation of tomato samples was carried out based on the conditions of their use for the production of tomato paste.

4.3.1 Material preparation

Tomato pulp, pomace and seeds residue were collected from recognized tomato paste industries in August and September 2020. They were cut into small pieces (1–0.5 cm). The tomato pomace and waste were high-moisture products, often more than 80% moisture and sometimes up to 98%, and they could spoil very quickly, in less than 2 days in some cases. Therefore, they were prepared immediately after manufacturing processing and were allowed to dry naturally at room temperature for 18 h, before being dried for 48 h at 50°C in order to be stabilized and stop enzymatic reactions.

Thereafter, the samples were subjected to a crusher to reduce particle sizes and then dried again for 4 hours at 65°C. The tomato residue samples were used during the bio-adsorption study in their natural state with no chemical or thermal treatment.

4.3.2 Estimation of heavy metals in effluents

All the chemicals used in this work were of analytical grade, and actual pharmaceutical effluent and wastewater samples were collected in eight groups of 5 L bottles from chemical laboratories (Inorganic Chemistry, Analytical Chemistry, General Chemistry, Organic Chemistry) within the Faculty of Pharmaceutical Chemistry. After collection, the wastewater effluent was immediately transported to the main research laboratory for analysis. Physico-chemical parameters such as pH, dissolved oxygen, total solids, total hardness, electrical conductivity, total dissolved solids, chloride, sulfate, calcium, sodium, cadmium, lead, nickel, zinc, copper, chrome, manganese, iron and potassium were analyzed as per the standard methods (AOAC, 2000; Arabian *et al.*, 2020; Shahsavan-Davoudi & Ziarati, 2020; Zahirnejad *et al.*, 2017). The synthetic Sigma-Aldrich stock solutions of Ni^{2+} and Co^{2+} (1000 mg/L) were used for making standard solutions. The desired concentrations of six standard solutions for each metal were prepared by successive dilutions of the stock solution.

A total of 20 mL of each wastewater sample was added to 3.0 mL of concentrated HNO_3 (65%, Merck) in a Pyrex beaker which was then covered with a watch glass and allowed to stand overnight. The covered beaker was then placed on a hot plate and heated at 95°C for 1 hour. Then the digestion solution was removed and left to cool. To prevent enzymatic reactions due to the organic compounds of tomato, 1 to 2 mL of H_2O_2 30% (Merck) was added to each digestion solution, as the preferred oxidant to avoid eventual precipitation problems without affecting the analyses (Ziarati & Ghasemynezhad-Shanderman, 2014).

All recoveries of the metals observed exceeded 95%. Ni and Co ion concentrations were determined in three replications by means of a Varian Vista inductively coupled plasma mass spectrometry (ICP-MS) device. The intra-day (for samples collected during the same day) and inter-day (for samples collected during different days) precision and accuracy of the method were determined under the optimal working conditions by triplicate measurements of known Ni/Co concentrations. The first standard stock solutions showed a 100.0 mg/L concentration of nickel (II) nitrate, and cobalt (II) nitrate, and aqueous standard solutions were prepared from them by means of appropriate dilution with 10% nitric acid. To determine nickel and cobalt concentrations by ICP-MS after specific times, the wet digestion method was used (Shahsavan-Davoudi & Ziarati, 2020; Zahirnejad *et al.* 2017). The samples were analyzed using a NexION 300X ICP-MS (Perkin-Elmer, USA). The instrumental operating conditions for the determination of the elements were set as per the analytical official methods (Tajik *et al.*, 2020).

4.3.3 Adsorbent particle size

The crusher sieve was used for the reduction of particle size of the dried tomato samples and for particle size distribution. The sieves were mechanically vibrated for 15 minutes which was sufficient for separation to take place. The screens were subjected to weighing balance before and after the vibration to get the mass and size of the dried tomato particles retained on each sieve. The particle size range used in this study was 0.15 mm to 3 mm.

4.3.4 Biochemical oxygen demand (BOD)

A total of 15 samples were immediately processed after collection for the determination of initial oxygen, and incubated at 20°C for 5 days for the determination of BOD_5. All the experiments were carried out in triplicate and the relative standard deviation was less than 2%. The sorption experiments were conducted with adsorbent doses of 0.1, 0.2, 0.3, 0.4, 0.5, 0.6, 1 and 2% in one litre of pharmaceutical effluent. Unstirred wastewater samples were studied at 0, 48 and 72 h, and 1 week, while samples stirred with an agitation rate of 200 rpm were studied at adsorbent times of 72 h and 1 week, all at room temperature (25 ± 3°C). To study the effect of pH on sorption, the pH of the metal ion solution was adjusted to values in the range of 2–10 prior to the experiment. The Langmuir isotherms were obtained by equilibrating metal ion solutions of different adsorbent doses (1–20) mg/L with different times (0–1 week) at equilibrium pH and rpm with an initial metal concentration of 10 mg/L at room temperature.

4.3.5 Total dissolved solids (TDS)

The total solid concentration in waste effluent represents the colloidal form and dissolved species. The probable reason for the fluctuation of the total solids value and subsequently the value of dissolved solids is due to content collision of these colloidal particles. The rate of collision of the aggregated process is also influenced by the pH of these effluents (Arabian *et al.*, 2020; Zahirnejad *et al.*, 2017).

4.3.6 Chemical oxygen demand (COD)

The chemical oxygen demand (COD) test determines, the oxygen required for chemical oxidation of organic matter with the help of a strong chemical oxidant. The COD level is used to measure pollution of domestic and industrial waste. The waste is measured in terms of equality of oxygen required for oxidation of organic matter to produce CO_2 and water. It is a fact that all organic compounds, with a few exceptions, can be oxidizing agents under the acidic condition. COD assessment is useful in pinpointing toxic conditions and presence of biological-resistant substances. For COD determination, samples were preserved using H_2SO_4 and processed for COD determination after the entire sampling operation was complete (Arabian *et al.*, 2020; Shahsavan-Davoudi & Ziarati, 2020; Zahirnejad *et al.*, 2017).

4.3.7 Statistical analysis

All experiments were performed three times and one way analysis of variance (ANOVA) was used for data analysis to measure the variations of metal concentrations using SPSS 22.0 software (SPSS Inc, IBM, Chicago, IL).

The means were compared and evaluated by Duncan's multiple range tests at the 95% level and *p*-value less than 0.05. In this work ICP-Mass was used to study the reduction rate of nickel and cobalt in laboratory effluents when in contact with tomato pomace powder as bio-adsorbent during different contact durations. So the variables are as follows:

- Time: a dependent variable with a quantitatively continuous scale.
- Bio-adsorbent concentration: a dependent variable with a slightly continuous scale.
- Investigation on efficacy of biosorbent along with the presence of stirring and without stirring: a dependent variable with a nominal qualitative scale.
- pH: a dependent variable with a slightly quantitative scale.

4.4 RESULTS AND DISCUSSION

The chemical composition profile of the wastewater effluent of the educational laboratories in the Faculty of Pharmaceutical Chemistry before treatment by tomato pomace are shown in Table 4.4. The tabulated data is the mean of three samples.

It is noted that COD was very high which is normal for effluent of such chemistry laboratories because there are over 2000 undergraduate students studying pharmaceutical chemistry and applied chemistry as bachelor degree

Table 4.4 Characteristics of wastewater from chemical laboratories in the Faculty of Pharmaceutical Chemistry, before treatment.

Parameters	Concentration
pH	4.03
BOD_5 at 20.8°C (mg/L)	3106
COD (mg/L): chemical oxygen demand	6680
TSS (mg/L): total suspended solids	52
Total alkalinity as $CaCO_3$ (mg/L)	94
Lead (mg/L)	4.832
Nickel (mg/L)	183.54
Mercury (mg/L)	0.09
Zinc (mg/L)	14.64
Cobalt (mg/L)	73.45
Chromium (mg/L)	3.15
Chloride (mg/L)	8.659
Sulfide (mg/L)	1.032
Nitrate (mg/L)	8718

programs, along with 100 students in master and PhD degree programs in the same laboratories. The mean content of BOD from 10 sub-samples was 3106 mg/L and the mean value of COD was 6680 mg/L for the sub-samples of the analyzed effluents.

Such findings from the studied laboratories are due to laboratory projects, experiments, and wet/acid digestion methods that had acidic states contributing 50% of the total waste flow at 3700 mm^3/day with a mean pH of 4.03 as shown in Table 4.4. The combined Analytical and Inorganic chemistry laboratories waste had a pH of 4.7 (including acidic waste stream), whereas the pH of the waste of the educational and research laboratories on the whole was 3.37.

Some heavy metal contents, such as cadmium, in the wastewater were found to be much higher than the limits according to IS-3306 (1974). Most of the solids present were in a dissolved form, with practically no suspended solids. This investigation was carried out to determine the accumulation of cadmium, nickel and cobalt in tomato pomace waste (TPW) in the contaminated wastewater of the educational laboratories in the faculty center, and in the chemical and toxicology laboratories' pharmaceutical effluent and wastewater after treating with bio-adsorbent at different specific times and doses of adsorbents, with and without stirring.

All experiments were performed three times and SPSS-20 software was used to analyze the variance of the data. The means were compared and evaluated by Duncan's multiple range tests at the 95% level and $p < 0.05$. In all figures, SE stands for standard error, and the Latin lowercase letters indicate the significant difference between the data.

Results from Figure 4.5 reveal that tomato residue has significant potential to adsorb cobalt ($p < 0.05$) during 6 hours even with very low content of

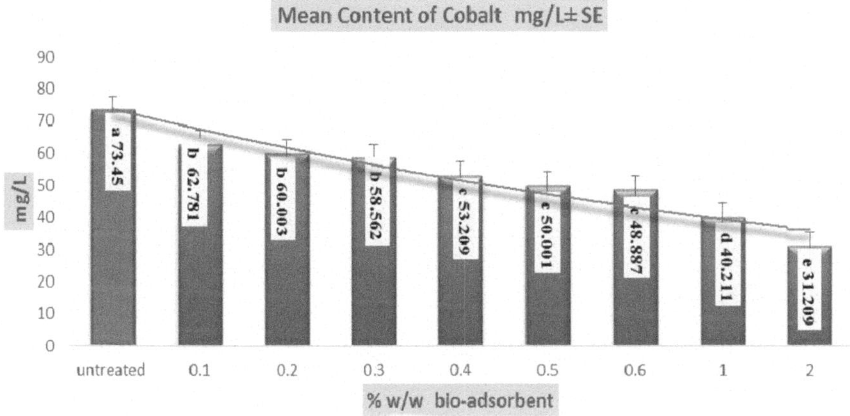

Figure 4.5 Effect of 6 hours contact time on the removal of cobalt from wastewater (initial Co concentration = 73.45 mg/L, bio-adsorbent doses = 0.1, 0.2, 0.3, 0.4, 0.5, 0.6, 1 and 2% tomato pomace waste, temperature = 25 ± 2°C).

bio-adsorbent (0.1% w/w). In the presence of 0.4, 1 and 2% (w/w) TPW, the adsorbing process increased significantly during the first hours of investigation and after 6 hours contact time in the presence of 2% TPW, reduction of Co was highly significant ($p < 0.001$) with the mean content decreasing from 73.45 ± 1.01 to 31.209 ± 0.78 mg/L ± SE. This suggests that the adsorption capacity varied by considering the effects of various parameters such as adsorbent dose.

The results of increasing the adsorbent to 0.5% (mg TPW/100 mL wastewater) in Figure 4.6, showed significant differences in cobalt up-taking by the bio-adsorbent after 1 week with and without the stirring factor. It was observed that increasing the contact time from 72 h to 1 week resulted in no significant

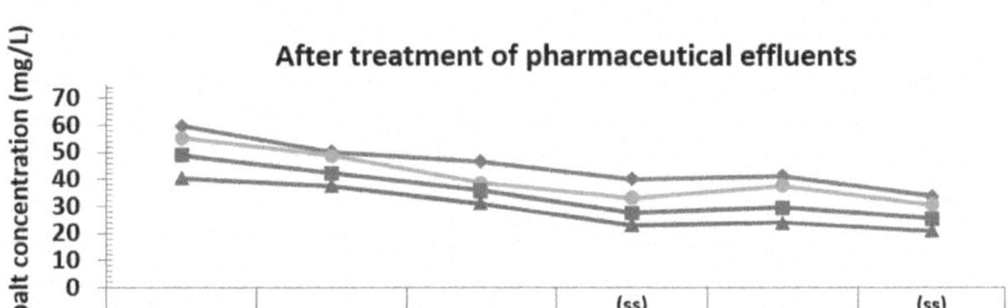

	6 h	24 h	72 h	(ss) 72 h	1 week	(ss) 1 week
0.1% bio-absorbent	59.66	50.12	46.5	40.04	41.21	33.81
0.3% bio-absorbent	55.3	48.77	38.72	33.2	37.6	30.7
0.5% bio-absorbent	48.9	42.2	35.8	27.8	29.8	25.7
1% bio-absorbent	40.55	37.65	31.2	23	24	21

Figure 4.6 Effect of contact time, with stirring factor on the removal of cobalt from wastewater (initial Co concentration = 73.45 mg/L, temperature = 25 ± 2°C, agitation speed = 200 rpm, pH = 4.0). ss: stands for stirred solution.

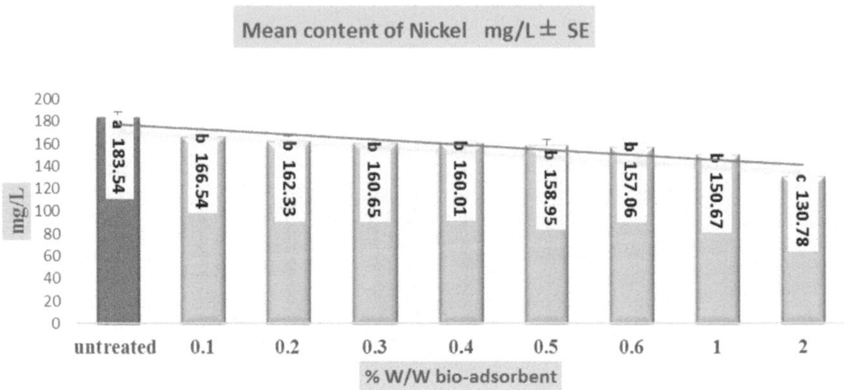

Figure 4.7 Effect of 6 h contact time on the removal of nickel from wastewater (initial Ni concentration = 183.54 mg/L, bio-adsorbent doses = 0.1, 0.2, 0.3, 0.4, 0.5, 0.6, 1 and 2% tomato pomace waste, temperature = 25 ± 2°C.

difference in heavy metal removal ($p \geq 0.05$). There was also no significant difference in removal between 1 week contact time with stirring and 2 weeks contact time without stirring (comparing data in Figures 4.4 and 4.5).

Results from Figure 4.7 reveal that tomato residue has significant potential to adsorb nickel ($p < 0.05$) during 6 hours even with very low content of bio-adsorbent (0.1% w/w). In the presence of 2% (w/w) further significant decreases were observed. The adsorbing process increased significantly during the first hours of investigation and after 6 hours contact time in the presence of 2% TPW, the reduction of Ni was significant ($p < 0.03$) with the mean content decreasing from 183.54 ± 1.00 to 130.78 ± 0.11 mg/L ± SE. This suggests the adsorption capacity for nickel is not as great as for cobalt.

All concentrations are expressed on the basis of mg/L ± SE. In Figure 4.8, ANOVA analysis showed that there was a significant difference ($p < 0.02$) in the concentration of nickel in the effluent of pharmaceutical laboratories with biosorbent of 1% TPW for different contact times (the difference between results from using 1% and 2% biosorbents was not significant). A dramatic decrease in Ni concentration is seen from 183.56 ± 1.14 to 76.51 (as in Figure 4.8) ± 0.09 mg/L ± SE.

The results in Figure 4.8 in the presence of 0.1% and 1% tomato pomace show almost the same results as for removal of cobalt but with a lower trend, and it can be concluded that this bio-adsorbent can accumulate selectively more cobalt than nickel ions with the same content of bio-adsorbent. Moreover, the study results showed that the contact time with the adsorbent, the percentage content of adsorbent, and the stirring factor all had significant effects ($p < 0.05$) on nickel removal.

This content of TPW adsorbent reduced the initial nickel concentration of 183.54 mg/L in untreated samples to 130.78 ± 1.01 mg/L ± SE ($p < 0.05$) after 48 h. After 2 weeks without stirring, the nickel content in the wastewater

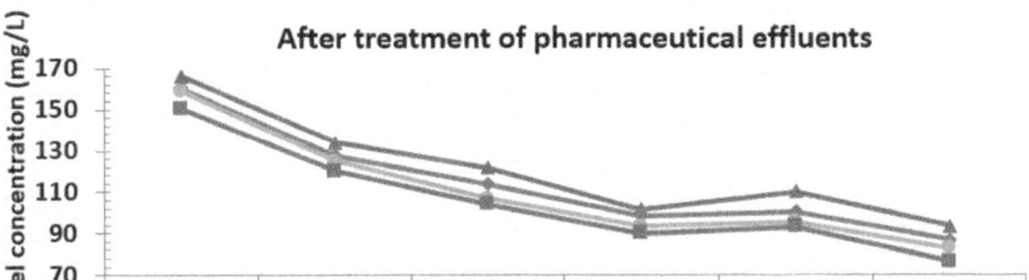

	6 h	24 h	72 h	(ss) 72 h	1 week	(ss) 1 week
0.1% bio-absorbent	166.54	134.28	121.71	101.54	110.09	93.48
0.3% bio-absorbent	160.65	127.65	113.72	98.67	100.28	86.82
0.5% bio-absorbent	158.95	125.03	106.89	93.29	95.44	82.68
1% bio-absorbent	150.67	120	103.92	90.01	93.01	76.51

Figure 4.8 Effect of contact time, with stirring factor on the removal of nickel from wastewater (initial Ni concentration = 183.54 mg/L, temperature = 25 ± 2°C, agitation speed = 200 rpm, pH = 4.0). ss: stands for stirred solution.

solution dropped dramatically to 66.21 ± 0.06 mg/L, which means removal of 63.93% of nickel, and in the presence of biomass along with the stirring factor it decreased to 25.414 mg/L in pharmaceutical effluent. In Figure 4.8, the removal trend in this green method can be seen.

The results in Figure 4.9 suggest that even with the lower concentration of tomato pomace food/agricultural waste of 0.1% and 0.2%, after only 6 hours the reduction process results are significant ($p < 0.05$), and after 24 and 72 hours cadmium content dropped significantly again ($p < 0.003$). The untreated wastewater contained 7.563 mg/L cadmium content, while after 2 weeks treating with 2% TPW and stirring the whole solution it decreased to 0.562 mg/L which shows 92.6% removal of this toxic elements without the use of any chemical compounds by an eco-friendly and green method.

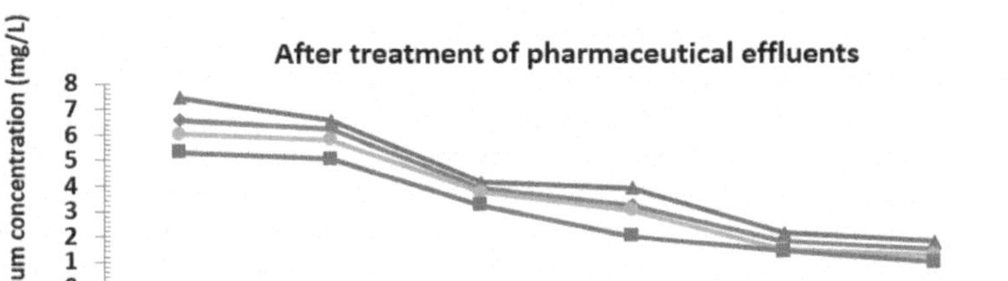

	6 h	24 h	72 h	(ss) 72 h	1 week	(ss) 1 week
0.1% bio-absorbent	7.44	6.55	4.11	3.89	2.11	1.78
0.3% bio-absorbent	6.56	6.22	3.88	3.21	1.78	1.46
0.5% bio-absorbent	6	5.78	3.71	3.01	1.45	1.21
1% bio-absorbent	5.3	5.04	3.21	2	1.38	0.94

Figure 4.9 Effect of contact time, with stirring factor on the removal of cadmium from wastewater (temperature = 25 ± 2°C, agitation speed = 200 rpm, pH = 4.0). ss: stands for stirred solution.

4.5 CONCLUSIONS

The use of biosorbents could be helpful in the development of an affordable 'green' environmental technology for purification of heavy-metals-contaminated drinking water which could present low income communities with the opportunity to benefit from heavy-metals-free drinking water to protect them from health hazards and toxins. The current study showed that by using TPW biomass, 92.6% of cadmium, 63.93% of nickel and 69.98% of cobalt can easily be removed from pharmaceutical-contaminated effluent.

Contamination by heavy metals is more prevalent due to inefficient food regulatory policies, and inadequate environmental monitoring and enforcement strategies. The health implications of trace elements and the toxic consequences of heavy metals necessitate effective monitoring of food products to ensure public health safety.

A decreasing trend in the level of Co content was achieved, even in contaminated wastewater from laboratories, by using TPW as a bio-adsorbent. It is suggested that the complicated functional group and molecules in TPW are rapidly excreted by chelating agents over a few hours or days leading to building a firm complex and the chelating agents–ligands which are able to form two or more coordinate covalent bonds with a cobalt metal ion. Further studies are required to increase understanding of the adsorption mechanism at the biosorbent–water interface.

REFERENCES

Alimardan M., Ziarati P. and Jafari Moghadam R. (2016). Adsorption of heavy metal ions from contaminated soil by B. integerrima barberry. *Biomedical & Pharmacology Journal*, **9**(1), 169–175, https://doi.org/10.13005/bpj/924

Amulya K., Dahiya S. and Venkata Mohan S. (2016). Chapter 19 – building a bio-based economy through waste remediation: innovation towards sustainable future. In: Bioremediation and Bioeconomy, M. N. V. Prasad (ed.), Elsevier, pp. 497–521. https://doi.org/10.1016/B978-0-12-802830-8.00019-8

AOAC (2000). Method 962.09, Official Methods of Analysis of AOAC International, 17th edn. 14th edn. AOAC International, Gaithersburg, Maryland USA.

Arabian S., Ziarati P. and Sawicka B. (2020). Waste herbal and black tea as a novel adsorbent for detoxification of pharmaceutical effluent. *Journal of Medical Discovery*, **5**(3), 1–15. https://doi.org/10.24262/jmd.5.3.20040

Arancon R., Lin C., Chan K., Kwan T. H. and Luque R. (2013). Advances on waste valorization: new horizons for a more sustainable society. *Energy Science & Engineering*, **1**, 53–71, https://doi.org/10.1002/ese3.9

Badr N. B., Al-Qahtani K. M. and Mahmoud A. E. D. (2020). Factorial experimental design for optimizing selenium sorption on Cyperus laevigatus biomass and green-synthesized nano-silver. *Alexandria Engineering Journal*, **59**(6), 5219–5229, https://doi.org/10.1016/j.aej.2020.09.051

Banerjee J., Singh R., Vijayaraghavan R., MacFarlane D., Patti A. F. and Arora A. (2017). Bioactives from fruit processing wastes: green approaches to valuable chemicals. *Food Chemistry*, **225**, 10–22. https://doi.org/10.1016/j.foodchem.2016.12.093

BLOG (2016). Udaleniye tyazhelykh metallov iz vody. Obzor. Ekologiya proizvodstva. http://www.ecoindustry.ru/user/skruber/blogview/2053.html (accessed 12 January 2022)

Caluya R. R., Sair R. R., Recta G. M. R. and Balneg B. B. (2003). Tomato pomace as feed for livestock and poultry, Mariano Marco State University. Available: http://research.mmsu.edu.ph/downloads/tomato_pomace.pdf (accessed 23 June 2007)

Del Valle M., Camara M. and Torija M. E. (2006). Chemical characterization of tomato pomace. *Journal of the Science of Food and Agriculture*, **86**, 1232–1236. https://doi.org/10.1002/jsfa.2474.74

Ehsani N., Ziarati P. and Salami M. (2021). Study on lead levels in infusion time of imported green and black tea (Camellia sinensis L). *Medical Science Journal of Islamic Azad University, Tehran Medical Branch*, **31**(2), 173–183.

El Din Mahmoud A. and Fawzy M. (2016). Bio-based methods for wastewater treatment: green sorbents. In: Phytoremediation: Management of Environmental Contaminants, A. A. Ansari, S. S. Gill, R. Gill, G. R. Lanza and L. Newman (eds.), Springer International Publishing, Cham, Vol. **3**, pp. 209–238.

Fatahi A., Ziarati P., Jafarpour A. and Cruz-Redrigouez L. (2020). Heavy metal removal from edible leafy vegetable by low cost novel adsorbents: hazelnut shell. *Journal of Science Discovery*, **4**(2), 1–13, https://doi.org/10.24262/jsd.4.2.20039

Fungaro D. A., Silva K. C. and Mahmoud A. E. D. (2021). Aluminium tertiary industry waste and ashes samples for development of zeolitic material synthesis. *Journal of Applied Materials and Technology*, **2**(2), 66–73, https://doi.org/10.31258/Jamt.2.2.66-73

Ganebnykh Y. V. (2010). Ochistka stochnykh vod ot ionov nikelya s ispol'zovaniyem gidrozoley montmorillonita. ZNiSO, 1. Available: https://cyberleninka.ru/article/n/ochistka-stochnyh-vod-ot-ionov-nikelya-s-ispolzovaniem-gidrozoley-montmorillonita (accessed 10 November 2021)

Garadaghli L.Ch. (2021). Influence of gray-brown soils contaminated with waste of sumgayit synthetic rubber plant on life activity of earthworms. *Advances in Biology & Earth Sciences*, **6**, 96–100. http://jomardpublishing.com/UploadFiles/Files/journals/ABES/V6N1/GaradaghliL.pdf (accessed 25 December 2021)

Heuzé V., Tran G., Hassoun P., Bastianelli D. and Lebas F. (2015). Tomato pomace, tomato skins and tomato seeds. Feedipedia, a programme by INRAE, CIRAD, AFZ and FAO. https://www.feedipedia.org/node/689 (accessed 12 December 2021)

Husain Khan A., Abdul Aziz H., Khan N. A., Ahmed S., Mehtab M. S. d., Vambol S., Vambol V., Changani F. and Islam S. (2020). Pharmaceuticals of emerging concern in hospital wastewater: removal of Ibuprofen and Ofloxacin drugs using MBBR method. *International Journal of Environmental Analytical Chemistry*, **2020**, https://doi.org/10.1080/03067319.2020.1855333

Husain Khan, A., Abdul Aziz, H., Khan, N.A., Ahmed, S., Mehtab, M.S., Vambol, S., Vambol, V., Changani, F., Islam, S. (2020a). Application of advanced oxidation processes followed by different treatment technologies for hospital wastewater treatment. *Journal of Cleaner Production*, **269**, 122411. https://doi.org/10.1016/j.jclepro.2020.122411

Jafari A., Amin G. and Ziarati P. (2016). Potential of echium ameonum fisch & mey in removing heavy metals from pharmaceutical effluent. *Bioscience & Biotechnology Research Asia*, **13**(3), 1585–1594, https://doi.org/10.13005/bbra/2303

Jallilian Z. and Ziarati P. (2016). High potential of Ferulago angulate (Schlecht) Boiss. in adsorption of heavy metals. *Biomedical and Pharmacology Journal*, **9**, 201–208, https://doi.org/10.13005/bpj/927

Khan N. A., Ullah Khan S., Ahmed S., Vambol S. and Vambol V. (2019a). Smart ways of hospital wastewater management, regulatory standards and conventional treatment techniques: a short review. *Smart and Sustainable Built Environmentthis Link is Disabled*, **9**(4), 727–736, https://doi.org/10.1108/SASBE-06-2019-0079

Khan N. A., Ahmed S., Vambol S., Vambol V. and Farooqi I. H. (2019b). Field hospital wastewater treatment scenario. *Ecological Questions*, **30**(3), 57–69, https://doi.org/10.12775/EQ.2019.022

Khan, N.A., Ahmed, S., Farooqi, I.H., Ali, I., Vambol, V., Changani, F., Yousefi, M., Vambol, S., Khan, S.U., Khan, A.H. (2020b). Occurrence, sources and conventional treatment techniques for various antibiotics present in hospital wastewaters: a critical review. *TrAC – Trends in Analytical Chemistry* **129**, 115921. https://doi.org/10.1016/j.trac.2020.115921

Khan N. A., Vambol V., Vambol S., Esrafili A. and Yousefi M. (2021). Hospital effluent guidelines and legislation scenario around the globe: a critical review. *Journal of Environmental Chemical Engineering*, **9**(5), 105874, https://doi.org/10.1016/j.jece.2021.105874

León-García E., Vela-Gutiérrezm G., Del Ángel-Coronel O. A., Torres-Palacios C., La Cruz-Medina J., Gómez-Lim M. A. and García H. S. (2017). Increased postharvest life of TomLox B silenced mutants of tomato (Solanum lycopersicum) Var. TA234. *Plant Foods for Human Nutrition*, **72**(4), 380–387. https://doi.org/10.1007/s11130-017-0629-y. PMID: 28918545

Mahmoud A. E. D. (2020a). Graphene-based nanomaterials for the removal of organic pollutants: insights into linear versus nonlinear mathematical models. *Journal of Environmental Management*, **270**, 110911. https://doi.org/10.1016/j.jenvman.2020.110911

Mahmoud A. E. D. (2020b). Eco-friendly reduction of graphene oxide via agricultural byproducts or aquatic macrophytes. *Materials Chemistry and Physics*, **253**, 123336, https://doi.org/10.1016/j.matchemphys.2020.123336

Mahmoud A. E. D. (2020). Nanomaterials: Green synthesis for water applications. In: Handbook of Nanomaterials and Nanocomposites for Energy and Environmental Applications, Kharissova O., Martínez L. and Kharisov B. (eds). Springer, Cham. https://doi.org/10.1007/978-3-030-11155-7_67-2

Mahmoud A. E. D. and Fawzy M. (2015). Statistical methodology for Cadmium (Cd (II)) removal from wastewater by different plant biomasses. *Journal of Bioremediation and Biodegradation*, **6**(304), 2.

Mahmoud A. E. D., Fawzy M. and Radwan A. (2016). Optimization of Cadmium (CD2+) removal from aqueous solutions by novel biosorbent. *International Journal of Phytoremediation*, **18**(6), 619–625. https://doi.org/10.1080/15226514.2015.1086305

Mahmoud A. E. D., Stolle A., Stelter M. and Braeutigam P. (2018). Adsorption technique for organic pollutants using different carbon materials. In: Abstracts of Papers of the American Chemical Society, Amer Chemical Soc, 1155 16th ST, NW, Washington, DC 20036 USA, Vol. **255**.

Mahmoud A. E. D., Fawzy M., Hosny G. and Obaid A. (2020). Equilibrium, kinetic, and diffusion models of chromium(VI) removal using Phragmites australis and Ziziphus spina-christi biomass. *International Journal of Environmental Science and Technology*, **18**(8), 2125–2136. https://doi.org/10.1007/s13762-020-02968-7

Mahmoud A. E. D., Umachandran K., Sawicka B. and Mtewa T. K. (2021a). Water resources security and management for sustainable communities. In: Phytochemistry, the Military and Health, A. G. Mtewa and C. Egbuna (eds.), Elsevier, pp. 509–522.

Mahmoud A. E. D., Al-Qahtani K. M., Alflaij S. O., Al-Qahtani S. F. and Alsamhan F. A. (2021b). Green copper oxide nanoparticles for lead, nickel, and cadmium removal from contaminated water. *Scientific Reports*, **11**(1), 12547. https://doi.org/10.1038/s41598-021-91093-7

Mehrarad F., Ziarati P. and Mousavi Z. (2016). Removing heavy metals from pharmaceutical effluent by plarganium grandiflorum. *Biomedical and Pharmacology Journal*, **9**, 151–161, https://doi.org/10.13005/bpj/922

Mousavi Z., Ziarati P., Esmaieli M. and Qomi M. (2014). Heavy metals (Lead and Cadmium) in some medicinal herbal products in Iranian market. *Iranian Journal of Toxicology*, **8**(24), 1004–1010.

Pimneva L. A. (2013). Ochistka stochnykh vod ot toksichnykh tyazhelykh metallov. *Sovremennyye Naukoyemkiye Tekhnologii*, **2**, 99–101.

Plazzotta S., Manzocco L. and Nicoli M. C. (2017). Fruit and vegetable waste management and the challenge of fresh-cut salad. *Trends Food Sci. Technol*, **63**, 51–59. https://doi.org/10.1016/j.tifs.2017.02.013

Rodionov A. I., Klushin V. N. and Sister V. G. (2000). Tekhnologicheskiye Protsessy Ekologicheskoy Bezopasnosti. Izd-vo N. Bochkarevoy, Kaluga, 800 s.

Sabiiti E. N. (2011). Utilising agricultural waste to enhance food security and conserve the environment. *African Journal of Food, Agriculture, Nutrition and Development*, **11**(6), 1–9.

Seifollahi R., Khoshkardar I. and Isvandi Z. (2017). Study on the removal of lead ions and nickel ions by natural agricultural biosorbents from industrial wastewaters (review study). 4th International Conference on Recent Innovations in Chemistry and Chemical Engineering, Tehran.

Shahsavan-Davoudi A. H. and Ziarati P. (2020). Green method for cadmium removal from pharmaceutical effluent laboratories by grapefruit peel. *Journal of Medical Discovery*, **5**(3). jmd20035. https://doi.org/10.24262/jmd.5.3.20035

Shirkhan F., Mostafidi M., Tamaskani Zahedi M., Ziarati P., Vambol V. and Vambol S. (2022). Green technologies and environmental management: a new understanding and approach to the use of agricultural waste. *Letters in Applied NanoBioScience*, **11**(1), 3065–3075. https://doi.org/10.33263/LIANBS111.30653075

Shokri F., Ziarati P. and Mousavi Z. (2016). Removal of selected heavy metals from pharmaceutical effluent by aloe vera L. *Biomedical and Pharmacology Journal*, **9**, 705–713, https://doi.org/10.13005/bpj/993

Tajik S., Ziarati P. and Cruz-Rodriguez L. (2020). Coffee waste as novel bio-adsorbent: detoxification of nickel from contaminated soil and coriandrum sativum methods. *Journal of Bioscience & Biomedical Engineering*, **1**(3), 1–11. https://doi.org/10.47485/2693-2504.1019

Tavakoli-Hosseinabady B., Ziarati P., Ballali E. and Umachandran K. (2018). Detoxification of heavy metals from leafy edible vegetables by agricultural waste: apricot pit shell. *Journal of Environmental & Analytical Toxicology*, **8**, 548. https://doi.org/10.4172/2161-0525.1000548

Valinurova E. R., Kadyrova A. D., Kudasheva F. K. H. and Sharafiyeva L. R. (2007). Ochistka vody ot ionov kobal'ta aktivirovannym uglerodnym voloknom. Izvestiya VUZov. *Chemistry and Chemical Technology*, **50**(9), 25–28. Available: https://cyberleninka.ru/article/n/ochistka-vody-ot-ionov-kobalta-aktivirovannym-uglerodnym-voloknom (accessed 10 October 2021)

Zahirnejad M., Ziarati P. and Asgarpanah J. (2017). The efficiency of bio-adsorption of heavy metals from pharmaceutical effluent by Rumex crispus L. Seed. *Journal of Pharmaceutical and Health Sciences*, **5**(3), 231–243.

Zakhvatkin V. V. (2012). Izvlecheniye kobal'ta i nikelya iz stochnykh vod. *Obogashcheniye rud*, **2**(2012), 43–44.

Ziarati P. (2012). Determination of contaminants in some Iranian popular herbal medicines. *Journal of Environmental & Analytical Toxicology*, **2**(1), 1, https://doi.org/10.4172/2161-0525.1000120

Ziarati P. and Ghasemynezhad-Shanderman S. S. (2014). Mineral contents in pleurotus (oyster mushroom): association of cooking method. *International Journal of Plant, Animal and Environmental Sciences*, **4**(2), 496–501.

Ziarati P. and Mirmaohammad Makki F. S. (2014). Lead, tin and cadmium contamination in canned tomato pastes. National Conference on Environment & Green Chemistry. Proceeding, 17–18 December 2014. University of Esfehan (6 pages).

Ziarati P. and Mirmaohammad Makki F. S. (2015). Removal of nitrate and nitrite from tomato derived products by lemon juice. *Biosciences Biotechnology Research Asia*, **12**(Spl. Edn. 2), 767–772, https://doi.org/10.13005/bbra/2258

Ziarati P. and Tosifi P. (2014). Comparing some physical and chemical properties of green olive (Olea Europea L.) in Iran association with ecological conditions. *International Journal of Plant, Animal and Environmental Sciences*, **4**(2), 519–528.

Ziarati P., Kermanshah A. and Moslehishad M. (2015). Adsorption heavy metal from contaminated water, by modified shell of wild endemic almonds: Amygdalus Lycioides and Amygdalus Wendelboi. *Biosciences Biotechnology Research Asia*, **12**, 2451–2457. https://doi.org/10.13005/bbra/1923

Ziarati P., Moslehishad M. and Mohammad-Makki F. M. (2016). Novel adsorption method for contaminated water by wild endemic almond: Amygdalus scoparia. *Biosciences Biotechnology Research Asia*, **13**, 147–153. https://doi.org/10.13005/bbra/2017

Ziarati P., Mostafidi M. and Arabian S. (2019a). Patent № 138281 Ukraina, Sposib ochystky vody. Zayavka u201904687.

Ziarati P., Mostafidi M., Arabian S., Vambol S., Vambol V. and Kozub P. (2019b). Patent № 137794 Ukraina, Sposib ochystky vody. Zayavka № u201903535; zayavl. 8.4.2019; opubl. 11.11.2019, Byul. № 22/2019.

Ziarati P., Farasati Far B., Mashayekhi E. and Sawicka B. (2019c). Removing arsenic by food-processing waste (Zizyphus jujuba seeds) and study on its adsorptive properties. *Technogenic and Ecological Safety*, **5**(1), 62–70. https://doi.org/10.5281/zenodo.2604648

Ziarati P., Kamyab-moghadas B. and Cruz-Rodriguez L. (2020a). Breast cancer disease and arsenic: geological behavior (GBR) as key. *Gynecology and Women's Health Care*, **2**(3), 1–8. https://unisciencepub.com/storage/2020/09/Breast-Cancer-Disease-And-Arsenic-Geological-Behavior-Gbr-As-Key.pdf (accessed 5 February 2022)

Ziarati P., Davoudi M. KH. SH., Kozub S., Vambol S., Vambol V. Kkhan N. A. and Akhmed S. (2020b). Sposib ochyshchennya vody [Method of water purification]. Ukrainian Patent NO. u 202002228, State Patent Office of Ukraine 2020, Kyiv.

Ziarati P., Davoudi M. KH. SH., Kozub S., Vambol S., Vambol V., Kkhan N. A. and Akhmed S. (2020c). Sposib ochyshchennya vody [Method of water purification]. Ukrainian Patent NO. u 2020 02190, State Patent Office of Ukraine 2020, Kyiv.

Ziarati P., Moradi D. and Vambol V. (2020d). Bioadsorption of heavy metals from the pharmaceutical effluents, contaminated soils and water by food and agricultural waste: a short review. *Labour Protection Problems in Ukraine*, **36**(2), 3–7. https://doi.org/10.36804/nndipbop.36-2.2020.3-7

Ziarati P., Kozub P., Vambol S., Vambol V., Khan N. A., Kozub S. and Tajik S. (2021). Kinetics of Cd, Co and Ni adsorption from wastewater using red and black tea leaf blend as a bio-adsorbent. *Ecological Questions*, **32**(2), 1–20. https://doi.org/10.12775/EQ.2021.014

Zverev M. P. (1981). Khemosorbtsionnyye Volokna. Khimiya, M., 192 s.

doi: 10.2166/9781789062625_0081

Chapter 5

Case studies of hospital effluent in urban treatment plants of other countries (USA/Canada)

Ana G. Pazmino-Sosa[1], Nishana P. Ramsawak[2] and Nastaran Mozaffari[3]*

[1]Centre Eau Terre Environnement, Institut national de la recherche scientifique (INRS), Quebec City, QC, Canada
[2]Civil Engineering Department, Queens University, Kingston, ON, Canada
[3]Department of Chemical Engineering, Universite Laval, Quebec, QC G1V 0A6, Canada
**Corresponding author. mozaffarinastaran7@gmail.com*

ABSTRACT

Hospital wastewater (HW), that is healthcare facilities effluents, is generated from different areas such as hospitalization, surgery, laboratories, laundries, and kitchens. It is discarded into the municipal sewage system, with or without previous treatment, increasing the pollutant concentrations to be treated in wastewater treatment plants. As a result, it is difficult to determine the exact contribution of pollutants from this source. HW is characterized in terms of its physical, chemical and biological properties, like pH, organic matter and nutrients concentration, conductivity, virus and bacteria presence, among others. HW has 2–3 times higher organic matter and suspended solids concentration compared to municipal wastewater, as well as higher presence of enteropathogenic viruses (Norovine, Adenovine, Rotavirus and Hepatitis A virus). Moreover, pharmaceutical compounds have been identified in HW such as analgesics, antidiabetics, antiepileptics, anti-inflammatories, diuretics, and psychoanaleptics. Further, this effluent is considered a crucial source of residual antibiotics in the environment that has been related to the existence and development of antimicrobial-resistant bacteria (ABR).

Keywords: Antibiotic, bacteria, health, hospital waste, pharmaceuticals, resistance

CONTENTS

5.1 INTRODUCTION

Healthcare facilities are different in size and service provided, and include general or specialized hospitals, isolation camps, burn patient units, feeding centers, public and private clinics, among others. These facilities are essential for population well-being and research development. The number of people that attend healthcare facilities daily varies according to the service offered, the location and the population income. For instance, it was estimated that in the USA 45 million procedures were performed and the total hospitalization rate increased from 1995 to 2010 according to the National Health Statistics Report (Hall *et al.*, 2017). As a result, unwanted by-products such as solid and liquid waste are generated, and they are treated according to the country-specific regulations (Verlicchi, 2018). With respect to the liquid waste, or hospital wastewater (HW), in some countries it is considered to be domestic effluent and treated accordingly, while in a few other countries, HW is handled as industrial effluent and is pre-treated before discharge into the municipal sewage system (Carraro *et al.*, 2016).

Because of its final disposal in sewage systems, it is difficult to differentiate the contribution of contaminants from HW into the treatment systems. However, this wastewater is recognized as one of the most polluted effluents, necessitating the determination of pollutant concentrations, their origin and fate. There are two methods available to do this: models development and measured concentrations. While the first method takes into account different parameters such as active ingredient consumption, water consumption, prescription rate, and excretion percentage, among others; measured concentrations involves sample collection and in-laboratory analysis (Verlicchi, 2018).

Nonetheless, to characterize any effluent, physico-chemical and biological parameters are relevant, and for HW, measurements of electrical conductivity (EC), biochemical oxygen demand (BOD), chemical oxygen demand (COD), total suspended solids (TSS), total nitrogen and phosphorus concentration, pH, fecal coliforms, pathogens and infectious agents are taken. Moreover, HW is distinctively characterized from other wastewaters due to its pharmaceuticals and antibiotics presence, like ciprofloxacin, spiramycin, clarithromycin, azithromycin and ofloxacin (Al Aukidy *et al.*, 2014), however, their values fluctuate within healthcare facilities (Aukidym, 2014) as well as seasonally (Aydin *et al.*, 2019).

This chapter aims to summarize the available information regarding the commonly found pollutants in HW and its physico-chemical and biological characteristics.

5.2 POLLUTANT LOAD ASSOCIATED WITH HOSPITAL WASTEWATER AND ITS ESTIMATION

Hospital wastewater (HW) refers to the wastewater produced by healthcare facilities from their different areas such as hospitalization, surgery, laboratories, laundries and kitchens (Verlicchi, 2018). It is estimated that the water consumption varies between 200 and 1200 L bed^{-1} day^{-1} in a hospital (Verlicchi *et al.*, 2013). This effluent is discarded, with or without previous treatment, in municipal sewage systems, and is then treated as urban effluent in wastewater treatment plants (WWTPs) (Al Aukidy *et al.*, 2014). HW is known to be an important source of pharmaceuticals, antibiotics, pathogens, and radioactive and toxic elements (Aydin *et al.*, 2019; Emmanuel *et al.*, 2002). It may also contain active substances of medicines and their metabolites, chemicals, heavy metals, disinfectants and sterilizers (Verlicchi *et al.*, 2013). According to Le Corre *et al.* (2012), HW could be the main source of antibiotics, such as ciprofloxacin, spiramycin, clarithromycin, azithromycin and ofloxacin, in municipal wastewater, as a result of their frequent prescription and consumption (Le Corre *et al.*, 2012). Moreover, 18% of the total quantity of the 'top 100 most-sold active compounds list of pharmaceuticals' in Switzerland is administered in hospitals (Escher *et al.*, 2011). On the other hand, Ort *et al.* (2010) compared the results obtained in a sampling campaign and the available data of pharmaceuticals consumption in Australia, determining that less than 15% of pharmaceuticals in wastewater came from HW. However, understanding the contribution of hospital and municipal wastewater separately is difficult. Therefore, there are methods available to make this differentiation possible: predicted and measured concentrations. To obtain predicted concentrations, modeling is an extensively used tool and it is principally based on parameters such as active ingredient consumption, water consumption per bed, excretion percentage, and half-life of the compounds; while sample collection and in-laboratory analysis are necessary to determine measured concentrations (Verlicchi, 2018). These methods have associated advantages and disadvantages, however, their accuracy highly depends on various parameters such as access to consumption information, access to sewage systems, the source of sampling, and the sample variability, among others, which could influence the results obtained (Verlicchi, 2018).

5.2.1 Physico-chemical and biological composition

In an initial step to understand the contribution of HW, research has characterized these effluents in terms of their physical, chemical and biological properties. The common values obtained are summarized in Table 5.1.

Different parameters are tested to identify the pollutants in wastewater, such as electrical conductivity (EC), biochemical oxygen demand (BOD), chemical oxygen demand (COD), total suspended solids (TSS), total nitrogen and phosphorus concentration, and pH. Moreover, when the wastewater source is an industry or hospital, other parameters are included such as total and free chlorine, detergents, disinfectants, oil and grease, tensoactives, sulphates, and microbiological indicators. In general, HW has 2–3 times higher contribution per patient for the concentration of BOD, COD and TSS compared with municipal wastewater (Verlicchi, 2018).

Biological characterization of HW includes the identification of fecal coliforms, pathogens and infectious agents. Regarding fecal coliforms, the analysis of *E. coli*, a predominant bacteria specie in the gut and feces, helps to indicate fecal contamination and the presence of other pathogenic fecal microorganisms (Harris *et al.*, 2014). Nonetheless, these bacteria are often found in greater quantities in municipal wastewater compared to HW because of the

Table 5.1 Hospital wastewater characterization parameters (Verlicchi, 2018; Verlicchi *et al.*, 2012).

Category	Parameter	Units	Value
Physico-chemical	Electrical conductivity	μS cm^{-1}	300–2700
	pH		6–9
	Redox potential	mV	850–950
	Fat and oil	mg L^{-1}	50–210
	Chlorides		80–400
	Total nitrogen		60–230
	Ammonium		10–68
	Nitrite		0.1–0.6
	Nitrate		1–2
	Phosphate		6–19
	Total suspended solids		116–3260
	Chemical oxygen demand		39–7764
	Total organic carbon		31–180
	Biochemical oxygen demand		16–2575
	Total surfactants		4–8
	Gadolinium	μg L^{-1}	<1–5
	Mercury		0.3–8
	Platinum		0.01–762
	Silver		$150–437 \times 10^3$
	Arsenic		0.8–11
	Copper		50–230
	Nickel		7–71
	Lead		3–19
	Zinc		70–670
Biological	*E. coli*	MPN 100 mL^{-1}	10^3–10^6
	Enterococci		10^3–10^6
	Fecal coliform		10^3–10^4
	Total coliforms		10^7–10^7
	Norovirus	Genomic copies L^{-1}	$2.4 - 10^6$
	Adenovirus		2.8×10^6
	Rotavirus		$1.9 - 10^6$
	Hepatitis A virus		104

intensive use of chemical disinfectants and antibiotics in healthcare facilities (Verlicchi *et al.*, 2012). On the contrary, concentrations of enteropathogenic viruses (Norovine, Adenovine, Rotavirus and Hepatitis A virus) in HW are 2–3 times greater than in municipal wastewaters, possibly related to the inpatients and diseases treatment (Carraro *et al.*, 2016).

Regarding heavy metals concentration, gadolinium (Gd), mercury (Hg) and platinum (Pt) are commonly found in HW (Verlicchi *et al.*, 2013). Gd is used as contrast media during magnetic resonance imaging (MRI), and it is excreted in a non-metabolized state in sewage systems, leading to a concentration range of 0.04–5 $\mu g\ L^{-1}$ in HW (Kümmerer, 2001). Hg is principally used in diagnostic and diuretic agents, and as amalgam as a restorative material in dental clinics (Singh *et al.*, 2014). The concentration in HW varies between 0.3 and 8 $\mu g\ L^{-1}$ (Verlicchi, 2018). Pt is added as an antineoplastic for oncological treatment and as an antitumor pharmaceutical (Lenz *et al.*, 2005). Moreover, platinum complexes are widely used to treat cancer diseases and are administrated to patients by an intravascular route with saline or glucose solution. It is excreted in a non-metabolized state, leading to concentrations of 0.01–762 $\mu g\ L^{-1}$ in HW (Dobrzynska *et al.*, 2021; Verlicchi, 2018).

5.2.2 Pharmaceutical composition

North Americans are considered the greatest producers and consumers of pharmaceuticals worldwide, representing the largest fraction of the total pharmaceutical revenues in this industry (Robinson *et al.*, 2007). The largest pharmaceutical market is driven by musculoskeletal drugs, followed by cardiovascular, oncological and anti-infective drugs; while the cholesterol-lowering drugs are the most consumed, followed by antidepressant, antihypertensive and antidiabetic drugs (González Peña *et al.*, 2021). The pharmaceuticals, are excreted mainly via urine and feces; however, disposal of unwanted leftover medications, non-metabolized substances, or conjugated metabolites with inactivating substances constitute a secondary route to their load in the environment (Daughton & Ruhoy, 2009). In HW, the range of pharmaceutical concentrations varies from several ng L^{-1} to mg L^{-1}, and is dependent on factors such as administered quantity, excreted percentage, and chemical characteristic of the compounds, among others (Ort *et al.*, 2010). HW is considered an important source of residual antibiotics such as ciprofloxacin, spiramycin, clarithromycin, azithromycin and ofloxacin (Al Aukidy *et al.*, 2014), however, this varies between healthcare facilities (Aukidym, 2014). In addition, Aydin *et al.* (2019) suggested that pharmaceutical concentrations in hospital effluents varied between summer and winter.

Pharmaceuticals are classified into more than 20 groups: analgesics, antibiotics, antibacterial, anti-infective, antidiabetics, antiepileptics, antihistamines, anti-hypertensives, anti-inflammatories, antifungals, anti-obesity, antithrombotic, calcium channel blockers, cardiac drugs, disinfectants, diuretics, lipid modifying agents, nasal decongestants, psychoanaleptics, psycholeptics, antagonists, steroid hormones, ß-agonists and ß-blockers (Oliveira *et al.*, 2015; Verlicchi *et al.*, 2013; Wiest *et al.*, 2018). However, determining the exact contribution of HW (containing pharmaceuticals) to municipal sewage systems is challenging because pharmaceuticals are used inside and outside the healthcare facilities. For example, antibiotics, analgesics, anti-neoplastics are mostly administered to inpatients, while lipid regulators or anti-hypertensives pharmaceuticals are given to outpatients for short or long periods of time (Verlicchi *et al.*, 2013).

Research performed by Oliveira *et al.* (2015) evaluated samples from hospital effluents to understand their impact on WWTPs in four water systems in Suffolk County, New York. The results showed that analgesics (22–26%), antidiabetics (12–31%), antiepileptics (7–15%), anti-inflammatories (4–10%), diuretics (3–8%) and psychoanaleptics (10–28%) composed 90% of the total average pharmaceuticals concentration in HW. From these groups, acetaminophen, gabapentin, ibuprofen, metformin, naproxen, theobromine and theophylline showed higher values than the average measured in all HW samples. It was also determined that between 1 and 59% of pharmaceuticals came from HW, but this varied between water systems sampled.

In Canada, Kleywegt *et al.* (2016) evaluated the presence of 141 pharmaceutical active compounds in healthcare facilities wastewater (two hospitals and a long-term care facility) and their contribution to the municipal sewage system in the province of Ontario. Their results showed a broad presence of pharmaceutical compounds in healthcare facility wastewaters, reaching concentrations at the magnitude of mg L^{-1}. Moreover, only 15 compounds from the 141 analyzed accounted for more than 1% of the total concentration of pharmaceuticals from healthcare facilities in the receiving WWTP. From these, the recurrent antibiotics identified were azithromycin, clarithromycin, ciprofloxacin, erythromycin, ofloxacin, sulfamethoxazole, and the antibacterial triclosan, while X-ray contrast media and cytotoxic drugs presence were rare. In addition, they determined that compounds from the groups of antibiotics or antimicrobial agents exceed environmental criteria.

As a result of residual antibiotics in the environment, the scientific community has raised the concern about the development and proliferation of bacterial resistance to antibiotics, pointing to the wastewater and its treatment facilities as potential reservoirs of antibiotic resistance bacteria (ARB) (Carraro *et al.*, 2016; Varela *et al.*, 2013). For example, in a waste stabilization pond (WSP) in the Clyde River (Canada), the effluent samples contained higher relative abundance of multiple antibiotic resistance genes than the influent, suggesting a microbial community enrichment during the treatment period (Ebrahim *et al.*, 2016). Regarding HW, it is expected to have greater presence and contribution of pathogens in treatment facilities due to the number of patients and bacterial diseases treated, however, HW is not the main influent in a treatment plant; therefore, it has little impact on the influent bacterial diversity (Verburg *et al.*, 2021). On the other hand, Kraupner *et al.* (2021) determined that untreated HW and residual antibiotics could drive and favor the evolution of resistance bacteria, concluding that HW serves as growth media for resistance strains, while residual antibiotics kill the susceptible bacteria in the media.

Moreover, a study performed by Edelsberg *et al.* (2014) evaluated the presence of common resistant pathogens isolated from 19 hospitals in the USA, identifying high frequency of fluoroquinolone-resistant *E. coli*, raising the concern that if the presence and growth of ABR in the environment continues, fluoroquinolones may no longer be effective to treat bacterial infections. Meanwhile, in Canada, ABR have been studied in different systems, like waste stabilization ponds in the Canadian Artic (Neudorf *et al.*, 2017) or the drinking water system (Mi *et al.*, 2019). Overall, the antimicrobial-resistant organisms identified in Canadian hospitals in 2008, methicillin-resistant *Staphylococcus aureus* (MRSA), different subgroups of *Streptococcus pneumoniae*, multi-drug resistant *N. gonorrhoeae* (Ebrahim *et al.*, 2016), continue their evolution and resistance development to antimicrobials. However, these studies have been performed in patient's isolated samples, which makes the identification of ABR presence challenging.

ABR is a major public health concern, and between untreated and treated wastewaters, including HW and municipal effluents, and the uncontrolled use and discharge of antibiotics, it will continue to increase and spread. Moreover, the Organisation for Economic Cooperation and Development (OECD) predicted that in the next 30 years, 2.4 million people will die from ABR infections in Europe, North America and Australia (Hofer, 2019).

5.3 IMMEDIATE HEALTH RISK OF HOSPITAL EFFLUENTS

Hospital effluents have been discovered as a major contributor to pharmaceuticals in wastewater, acting as a micropollutant. These effluents are known to contain microorganisms as well as active pharmaceutical ingredients, in combination with each other. In recent times, the study of hospital effluents has increased, with focus on microbial activities in the wastewater. Some WWTPs are not specifically designed to remove these persistent active ingredients in pharmaceuticals and have been shown to pose human health risks, as accumulation of antibiotic resistance over time has evolved to affect humans as well as animals when released through hazardous wastewater during events such as flooding, and/or sewer overflow. The release of antibiotics has also led to the prevalence of antibiotic resistance genes in the environment, causing the emergence and spread of resistant bacteria. These pollutants have been found to cause fish death and affect algae at even low doses. The uncontrolled discharge of effluent into the environment influences quite a few waterborne diseases and poses a significant threat to human life. Hospital effluents are known to contain radioactive isotopes, pharmaceuticals, stock cultures, heavy metals, pathogens, drugs, cotton particles, disinfectants, and chemical compounds (Ahsan, 2012). Hospital wastes have been long associated with contraceptive-rich pharmaceuticals which act as endocrine disruptors, which, have in the past been linked to reproductive impairment in fishes (Obasi *et al.*, 2014), eutrophication due to oxygen depletion, algal bloom, and death of aquatic animals if not properly treated (Buelow *et al.*, 2017). Akin (2016) reported the average discharge of hospital wastewater depends on the size and capacity of the hospital and can range from 400 to 1200 L day^{-1} bed^{-1}. The immediate health concerns regarding hospital effluents are for the environment and human health as they contain infectious bacteria such as prions, viroid, and toxins (Kaur *et al.*, 2020). The pharmaceuticals in hospital effluents have been reported to undergo biological alterations, forming conjugated complexes, resulting in a greater number of toxicological compounds being present than in the unconjugated metabolites (Tiwari *et al.*, 2017). These complexes are present in low concentrations and are not fully removed during pre-treatment or treatment of these effluents at wastewater treatment plants (Verlicchi *et al.*, 2010). Hospital effluents provide the ideal environment for microbes such as bacteria, viruses, fungi, and parasites to thrive. The effluents also contain several strains of antibiotic residues and antibiotic-resistant bacteria, which stifles the growth rate of susceptible bacteria, allowing for increased microbial-resistant bacteria to be released further downstream from the treatment plant, which further poses a threat to public health (Asfaw *et al.*, 2017). In work done by Efaq *et al.* (2015) it was found that the growth rate of fungi increases, which allows them to spread their spores to their surroundings, which also results in a negative consequence on the environment and human health. According to work done by Chonova *et al.* (2018), the genus Bacillus was the most predominant bacteria in hospital wastewater accounting for 80 to 90% of the total, with *Staphylococcus* and *Streptococcus* accounting for 5 to 10%. Hospital effluent also contains microbes that are drug resistant such as Acinetobacter, enterococcus, and pseudomonas species (Chonova *et al.*, 2018; Kaur *et al.*, 2020).

Microbial analysis of hospital effluents has found the occurrence of fungus, bacterial and coliform species (Efaq *et al.*, 2017). While fungus can easily spore in the environment, nature, and the environmental factors such as pH, temperature, nutrients, and moisture promote the ideal environment for vast growth (Srinivasan *et al.*, 2014). The prevalence of Rotavirus A, norovirus geno I and II, human adenovirus and hepatitis was analyzed from samples collected in two different hospital wastewater treatment plants, and was found to be higher than the load detected in urban sewage plants (Prado *et al.*, 2011). Release of hospital effluents directly into the municipal wastewater without a pre-treatment process in a wastewater treatment plant, can increase the occurrence of gastroenteric viruses in water systems (Ibrahim *et al.*, 2018) and also increases the potential for skin diseases and other infections dangerous to human health (Okjokwu & Inabo, 2014). The potential dangers of hospital effluents are becoming more of a concern and require specific and advanced technologies, which can be site-specific for optimal protection of environmental and human health. Some environmental impacts of hospital effluents include release of excessive nutrients which contributes to biological degradation in groundwater, lakes, and rivers by eutrophication, causing algal bloom and biotoxins. Some waterborne diseases that can result from hospital effluents are campylobacteria, cholera, hepatitis A and E, schistosomiasis and typhoid fever, but hospital effluents can also influence vector-borne diseases and parasites such as dengue fever, malaria, and round worms (Stewart *et al.*, 2011). Additionally, when users take drugs and some remain non-metabolized or only partially assimilate, these pharmaceuticals can be excreted and end up in municipal wastewater. The pathogens from hospital effluents can survive for a lengthy period in water and soil and make their way into the food chain, causing infectious diseases and further health risks to humans (Malakar *et al.*, 2019). While the direct effect of hospital effluents remains a widely discussed emerging topic in literature, it is necessary to work now on limiting the spread and magnitude of potential threats to human health and the environment. Despite the many potential health risks associated with hospital effluents, very few studies on this topic exist, particularly with emphasis on viral diseases and pathogenic spread (Akin, 2016).

Table 5.2 shows some of the major bacterial isolates found in hospital environments (Kaur *et al.*, 2020).

Table 5.2 Some of the major bacterial isolates found in hospital environments.

Bacterial Isolate	Identification Methods	Source	Reference
Enterobacteriaceae or Aeromonas, Kluyvera	Matrix assisted laser desorption/ionization-time of flight mass spectroscopy and gene sequencing	Ciprofloxacin, cotrimoxazole, chloramphenicol, cephalosporins, fluoroquinolones, and aminoglycosides	Picão *et al.* (2013)
Carbapenem-resistant Enterobacteriaceae	Microbiological methods and quantitative polymerase chain reaction (qPCR)	Carbapenem combinations, polymyxins, fosfomycin, ceftazidime-avibactam	Lamba and Ahammad (2017)
E. coli, Salmonella, Shigella, S. aureus	Biochemical and microbial tests	Ioperamide, trimethoprim sulfamethoxazole, cefotaxime, β-lactams, quinolones, macrolides, nafcillin, sulfa drugs, and vancomycin	Fedaku *et al.* (2015)
Enterococci	Microbiological methods, PCR	Piperacillin tazobactam, imipenem, meropenem, ampicillin	Iweriebor *et al.* (2015)
Extended-spectrum beta-lactamase (ESBL)-producing Enterobacteriaceae	Microbiological and 16 rRNA sequencing	Carbapenem, cephamycin, nitrofurantoin, clavulanic acid, and tazobactam	Korzeniewska and Harnisz (2013)
Fecal bacterium, Aeromonas, Prevotella	16S rRNA-based denaturing gradient gel electrophoresis analysis	Chloramphenicol, ciprofloxacin, cotrimoxazole, aminoglycosides, metronidazole, ureidopenicillin, carbapenems, clindamycin and pavilion	Varela *et al.* (2013)
Klebsiella, Pseudomonas, Serratia	Microbiological and molecular methods	Ampicillin, sulbactam, ceftazidime, cefepime, levofloxacin, norfloxacin, gentamycin, aztreonam, aminoglycoside with antipseudomonal β-lactam	Ekhaise and Omavwoya (2008)

(Continued)

Table 5.2 Some of the major bacterial isolates found in hospital environments (*Continued*).

Bacterial Isolate	Identification Methods	Source	Reference
Klebsiella pneumoniae carbapenemase (KPC)-producing *Aeromonas* spp., Enterobacteriaceae, and ESBL-producing *K. pneumonia, Enterobacter, E. coli*	Microbiological, PCR, and 16S rRNA sequencing	Chloramphenicol, ciprofloxacin, cotrimoxazole, cephalosporins, fluoroquinolones, ampicillin, ticarcillin/clavulanate, ertapenem, and meropenem	Chagas *et al.* (2011); Picão *et al.* (2013)
E. coli, S. aureus	Microbiological and molecular methods, polymorphic DNA-PCR	Isothiocyanates, cephalosporins, nafcillin, trimethoprim, nitrofurantoin, sulfa drugs, pivmecillinam, vancomycin	Galvin *et al.* (2010); Katouli *et al.* (2012); Lien *et al.* (2017)
Methicillin-resistant *S. aureus*, vancomycin-resistant *S. aureus*	Biochemical tests, PCR, and sequencing	Doxycycline, minocycline, trimethoprim, sulfamethoxazole, nitrofurantoin, linezolid daptomycin, and tigecycline	Mandal *et al.* (2015)
Proteobacteria, Bacteroidetes, Firmicutes, Actinobacteria	Microbiological, qPCR, and 16S rRNA sequencing	Azithromycin, doxycycline, carbapenem, tigecycline, cefoxitin, and penicillin G	Szekeres *et al.* (2017)
Streptococcus, Achromobactin, *Ruminococcaceae*	16S rRNA sequencing, and qPCR	Meropenem, imipenem, trimethoprim-sulfamethoxazole, ticarcillin, and cephalosporins	Beulow *et al.* (2017)

5.4 FULL SCALE HOSPITAL EFFLUENT TREATMENT – CASE STUDY

Hospital wastewater has similar characteristics as municipal wastewater plus the presence of different pollutants such as pharmaceuticals, heavy metals, disinfectants, pathogens and bacteria (Verlicchi *et al.*, 2012). Commonly, this effluent is treated jointly with municipal wastewater in treatment facilities (WWTPs). Normally, a WWTP includes primary, secondary and disinfection treatment before discharge. At the primary or physical reactor, total suspended solids, organic matter and phosphorus treatment occurs, while at the secondary or biological reactor, ammonium and remaining organic matter are treated. Finally, the wastewater is disinfected with UV light or chlorine addition to remove bacteria and viruses before final disposal to aquatic bodies (Tchbanoglous *et al.*, 2003). The efficiency of this technology is over 88% for organic matter removal, as well as ammonium and phosphorus (Zhang *et al.*, 2016). However, these facilities are not equipped to treat pharmaceuticals, antibiotics or emerging contaminants present in municipal and hospital wastewater. Moreover, there are no available guidelines for discharge of these contaminants (Rodriguez-Mozaz *et al.*, 2017). To date, the contribution of hospitals to the pharmaceutical load in wastewater remains unclear, however, the use of alternative treatment technologies for HW have been suggested (Escher *et al.*, 2011).

According to a research review by Verlicchi (2018), a total of 23 studies reported full-scale treatment for HW between 2004 and 2016, mentioning places like Brazil, China, Germany, Denmark, Greece, Italy, Iran, Taiwan, Korea, Ethiopia, Saudi Arabia, India, Nepal, and Vietnam. In the USA, numerous potentially hazardous and xenobiotic organic chemicals found in wastewater have been detected downstream of wastewater sites, with over 100 wastewater chemically derived organics being identified by Facazio *et al.* (2008) in the USA alone. Cooper *et al.* (2008) designed a database as a preliminary risk assessment for pharmaceuticals found in wastewater effluents. Drugs that act on the central nervous system, and several antibiotics were found to pose greater risk to human health in these effluents (Daughton & Ruhoy, 2008). While many reports have indicated the adverse potential health effects, some reports have also reported negligible effects. In Michigan, near the Saginaw and Shiawassee rivers and Saginaw Bay, the toxicity levels in the watersheds and wastewater treatment plants were examined. Kannan *et al.* (2008) found minimal concerns having reported a metamizole residue at 7 $\mu g\ L^{-1}$ in wastewaters, which is not of concern for humans or animals.

Improperly treated or mismanaged wastewater, in particular hospital effluents released into water systems without efficient removal capabilities, poses a serious threat to humans and the environment. In a study conducted from 2000 to 2004 in Southern California, Brinks *et al.* (2008) estimated 689 000 to 4 003 000 gastrointestinal illness occurrences and 693 000 respiratory illness occurrences happened each year from exposure to beaches where effluent discharged into the coastal waters from smaller clinics and industries. In New Jersey, Donovan *et al.* (2008) evaluated the Lower Passaic River, as a result of combined sewer overflow, for gastric illness and Giardia infection and found the river water had a high concentration of fecal pathogens.

Kostich and Lazorcak (2008) investigated the risk to aquatic life associated with exposure to prescription pharmaceutical residues from effluents in the USA and reported that some active pharmaceutical ingredients were restrictive to just vertebrates while some affected just eukaryotic and prokaryotic clades. Hospital effluents are rich in active pharmaceutical ingredients, and some are not thoroughly removed and have been found to accumulate at even trace levels. The effect of hospital effluents from urban plants in North America has been found to vary based on season as well. Alvarez *et al.* (2008) investigated the impact of seasonality on effluents and their occurrences of organic contaminants, including atrazine, metolachlor, and polycyclic aromatic hydrocarbons as well as other chemicals. Iwanowicz *et al.* (2009) conducted a study in the Potomac River, USA on the reproductive health of bass fishes, and found that estrogenic endocrine disruptors in effluents had a greater influence on intersexuality, with a greater intersex prevalence being found in male bass. Thorpe *et al.* (2008) used fathead minnows to monitor the estrogenic effects of temperate effluents, replicable to other population endpoints. In the Red Deer River, Alberta, Canada, Jeffries *et al.* (2009) studied longnose dace (Rhinichthys cataractae) to monitor the changes in population, growth, and physiological performance endpoints. Jeffries found that those captured immediately downstream of the wastewater treatment plants had bigger liver and gonad sizes than those captured further away. Basmadjian *et al.* (2008) investigated spatial differences in liver lesions in other aquatic life at fixed locations in San Pedro shelf, California, USA where there is an effluent point, and found no consistent spatial differences in lesion prevalence.

From hospital effluents, microbial loads were reported to be, as expected, higher during the day as compared to evenings and nights which can be attributed to the fact that activities in hospitals are more restrictive at night than in the daytime. In hospital effluents, the dominant phylums identified are Proteobacteria and Firmicutes, correlated to their presence in human feces, long-term survival ability in wastewater, and their capacity to survive in extreme environmental contaminants levels (Li *et al.*, 2011). Hospital effluents also contain different organisms of human gut microbiota such as genus *Streptococcus* and *Ruminococcaceae*, however, they do not survive well in complex environments (Buelow *et al.*, 2018). In addition, presence of anaerobic bacteria genera such as Bifidobacterial, Clostridial, and Bacteroidales in hospital wastewater has been reported (Buelow *et al.*, 2017). Moreover, the release of antibiotics creates a selection pressure on bacteria resulting in a rapid adaptation to fluctuating conditions through genome rearrangement in innate organisms (Santoro *et al.*, 2015) leading to the prevalence and spread of antibiotic-resistance bacteria (ARB). A study identified *E. coli* from hospital effluent which was resistant toward ampicillin, streptomycin, sulfamethoxazole, cephalosporin, ciprofloxacin, and tetracycline (Galvin *et al.*, 2010). In addition to ARB, HW may also carry other pathogenic microorganisms such as viruses, fungi, protozoans, and helminths (Casas *et al.*, 2015; Iweriebor *et al.*, 2015; Santoro *et al.*, 2015).

Due to the diversity of contaminants, HW could be 5–15 times more toxic compared to urban effluents and its characteristics could inhibit the activated sludge population in treatment plants (Biobooster 2016; Boillot *et al.*, 2008; Watkinson *et al.*, 2009). Therefore, HW collection and treatment could minimize the potential risks for local populations (Iweriebor *et al.*, 2014; Timraz *et al.*, 2017). There is little information that correlates human health risks with pharmaceuticals in the environment in Canada's waterways, contrary to data availability in the USA or Europe, making it difficult to extrapolate (Khan & Nicell, 2015). For example, in 2014, 56 pharmaceuticals were identified in 50 large treatment plants, finding that every water sample contained the diuretic hydrochlorothiazide, while over 90% of the samples contained beta-blockers and carbamazepine. From these, carbamazepine, gemfibrozil, and ibuprofen were detected in Ontario (Kostich *et al.*, 2014). Moreover, (Atkinson *et al.*, 2012) identified in wastewater from a south-eastern Ontario treatment plant, the presence of estrone, 17β-estradiol and 17α-ethinylestradiol (EE2).

Regarding hospital and long-term care home effluent, in 2012 the first study conducted on Canadian effluent was released, where nine pharmaceuticals were analyzed to identify their occurrence, mass flows and day-to-day variability as well as the relative contribution of target pharmaceuticals by the hospital effluents. The findings suggested that healthcare facilities contributed a large proportion of the overall load of antibiotics in municipal wastewater treatment plants (Riaz ul Haq *et al.*, 2012, p. 141). Most of the hospitals in Ontario, Canada do not pre-treat effluents, but their wastes are sorted before disposal. The effluents are released into municipal wastewater and treated as part of municipal wastewater. The Water Quality Guideline of Pharmaceuticals developed in British Columbia is used as a guideline for regulations, but it is not strictly managed by each province in Canada. A study by Khan and Nicell (2015) intended to quantify the exposure of Canadians to pharmaceuticals based on predicted concentrations of pharmaceutically active compounds (PhACs) in Canada, derived from Canada-specific prescription data. The observed results showed that hospitals accounted for less than 10% of the total environmental load from all sources considered, yet hospitals were an important point source of antibiotics, corroborating findings described by Riaz ul Haq *et al.* Therefore, more research regarding the potential effects of pharmaceuticals in the environment and the risk to human health is needed. Also, better monitoring and characterization of HW should be performed to clearly identify its contribution to treatment plants, particularly with respect to antibiotics.

5.5 CONCLUSION

Health clinics and centers possess essential healthcare facilities in various sizes in order to support the well-being of people and research development. Hospital wastewater (HW) contains unwanted by-products such as solid and liquid phases of waste generated from healthcare facilities and their different areas such as hospitalization, surgery, laboratories, laundries and kitchens, and it is one of the important parts of wastewater which should be treated appropriately or pre-treated before discharging into the municipal sewage system. Since this wastewater is recognized as one of the most contaminated effluents, it is necessary to measure and characterize its electrical conductivity (EC), biochemical oxygen demand (BOD), chemical oxygen demand (COD), total suspended solids (TSS), total nitrogen and phosphorus concentration, pH, fecal coliforms, pathogens, heavy metals, and other agents.

In North America, HW contains pharmaceuticals in the range from several ng L^{-1} to mg L^{-1} depending on factors such as administered quantity, excreted percentage, chemical and biological properties, and also seasonal parameters. In a recent research study, Aydin *et al.* (2019) revealed the variation of pharmaceutical concentrations in HW effluents between summer and winter, especially when the weather is colder providing conditions for increased cases of influenza. Alvarez *et al.* (2008) also concluded that some organic contaminants such as atrazine, metolachlor, polycyclic aromatic hydrocarbons, and so on., appearing in HW effluents would be impacted by seasonal changes. Research conducted by Kleywegt *et al.* (2016), showed a broad presence of pharmaceutical compounds in healthcare facility wastewaters, reaching concentrations at the magnitude of mg L^{-1}, in HW effluent of Ontario, Canada, in which antibiotics or antimicrobial groups are dominant.

Research performed by Oliveira *et al.* (2015) evaluated samples from hospital effluents to understand their impact on WWTPs in four water systems in Suffolk County, New York. The results showed that analgesics (22–26%), antidiabetics (12–31%), antiepileptics (7–15%), anti-inflammatories (4–10%), diuretics (3–8%) and psychoanaleptics (10–28%) composed 90% of the total average pharmaceuticals concentration in HW. Other agents including acetaminophen, ibuprofen, naproxen, gabapentin, and so on., showed higher values than the average measured in HW samples. An investigation into the presence of common resistant pathogens present in 19 hospitals in the USA was conducted by Edelsberg *et al.* (2014), which revealed the high frequency of fluoroquinolone-resistant *E. coli* in HW samples.

Overall, the most commonly detected agents in the HW effluents of both Canada and the USA are pharmaceutical drugs such as antibiotics, bacterial agents such as *E. coli*, and heavy metals. Also, it is necessary to conduct further research on the effect of long-term and active agents on aquatic species, in cases where HW is discharged into the sewage or the environment without any pre-treatment or treatment.

REFERENCES

Ahsan N. (2012). Study of widely used treatment technologies for hospital wastewater and their comparative analysis. *International Journal of Advanced Engineering and Technology*, **5**(1), 227.

Akin B. S. (2016). Contaminant properties of hospital clinical laboratory wastewater: a physiochemical and microbiological assessment. *Journal of Environmental Protection*, **7**(05), 635, https://doi.org/10.4236/jep.2016.75057

Al Aukidy M., Verlicchi P. and Voulvoulis N. (2014). A framework for the assessment of the environmental risk posed by pharmaceuticals originating from hospital effluents. *Science of the Total Environment*, **493**, 54–64, https://doi.org/10.1016/j.scitotenv.2014.05.128

Alvarez D. A., Cranor W. L., Perkins S. D., Schroeder V. L., Iwanowicz L. R., Clark R. C., Guy C. P., Pinkney A. E., Blazer V. S. and Mullican J. E. (2009). Reproductive health of bass in the Potomac, USA, drainage: Part 2. Seasonal occurrence of persistent and emerging organic contaminants. *Environmental Toxicology and Chemistry: An International Journal*, **28**(5), 1084–1095. https://doi.org/10.1897/08-417.1

Asfaw T., Negash L., Kahsay A. and Weldu Y. (2017). Antibiotic resistant bacteria from treated and untreated hospital wastewater at Ayder Referral Hospital,

Mekelle, North Ethiopia. *Advances in Microbiology*, **7**, https://doi.org/10.4236/aim.2017.712067

Atkinson S., Marlatt V., Kimpe L., Lean D., Trudeau V. and Blais J. (2012). The occurrence of steroidal estrogens in south-eastern Ontario wastewater treatment plants. *Science of the Total Environment*, **430**, 119–125, https://doi.org/doi.org/10.1016/j.scitotenv.2012.04.069

Aydin S., Aydin M. E., Ulvi A. and Kilic H. (2019). Antibiotics in hospital effluents: occurrence, contribution to urban wastewater, removal in a wastewater treatment plant, and environmental risk assessment. *Environmental Science and Pollution Research*, **26**(1), 544–558, https://doi.org/10.1007/s11356-018-3563-0

Basmadjian E., Perkins E., Philips C., Heilprin D., Watts S., Diener D., Myers M., Koerner K., Mengel M., Robertson G., Armstrong J., Lissner A. and Frank V. (2008). Liver lesions in demersal fishes near a large ocean outfall on the San Pedro Shelf, California. *Environmental Monitoring and Assessment*, **138**, 239–253, https://doi.org/10.1007/s10661-007-9794-z

Biobooster G. (2016). Full scale advanced wastewater treatment at Herlev hospital treatment performance and evaluation. *Tech. Rep.*, May 2016. https://www.dhigroup.com/global/news/2016/08/hospital-wastewater-from-a-pollution-problemto-new-water-resources

Boillot C., Bazin C., Tissot-Guerraz F., Droguet J., Perraud M., Cetre J., Trepo D. and Perrodin Y. (2008). Daily physicochemical, microbiological and ecotoxicological fluctuations of a hospital effluent according to technical and care activities. *Science of the Total Environment*, **403**(1–3), 113–129, https://doi.org/10.1016/j.scitotenv.2008.04.037

Brinks M., Dwight R., Osgood N., Sharavanakumar G., Turbow D., El-Gohry M., Caplan J. and Semenza J. (2008). Health risk of bathing in southern California coastal waters. *Rchives of Environmental & Occupational Healt*, **63**(3), 123–135, https://doi.org/10.3200/AEOH.63.3.123-135

Buelow E., Bayjanov J., Williem R., Bonten M., Schmitt H. and van Schaik W. (2017). The microbiome and resistome of hospital sewage during passage through the community sewer system. *BioRxiv*, https://doi.org/doi.org/10.1101/216242

Buelow E., Bayjanov J., Majoor E., Williem R., Bonten M., Schmitt H. and van Schaik W. (2018). Limited influence of hospital wastewater on the microbiome and resistome of wastewater in a community sewerage system. *Microbiology Ecology*, **94**(7), https://doi.org/10.1093/femsec/fiy087

Carraro E., Bonetta S., Bertino C., Lorenzi E., Bonetta S. and Gilli G. (2016). Hospital effluents management: chemical, physical, microbiological risks and legislation in different countries. *Journal of Environmental Management*, **168**, 185–199, https://doi.org/10.1016/j.jenvman.2015.11.021

Casas M., Chhetri R., Ooi G., Hansen K., Litty K., Christensson M., Kragelund C., Andersen H. and Bester K. (2015). Biodegradation of pharmaceuticals in hospital wastewater by a hybrid biofilm and activated sludge system (Hybas). *Science of the Total Environment*, **530–531**, 383–392, https://doi.org/10.1016/j.scitotenv.2015.05.099

Chagas T. P., Seki L., Cury J., Oliveira J. A., Davila A. M., Silva S. and Asensi M. (2011). Multiresistance, beta-lactamase-encoding genes and bacterial diversity in hospital wastewater in Rio de Janeiro, Brazil. *Applied Microbiology*, **111**(3), 572–581.

Chonova T., Labanowski J., Cournoyer B., Chardon C., Keck F., Laurent E., Mondamert L., Vasselon V., Wiest L., & Bouchez, A. (2018). River biofilm community changes related to pharmaceutical loads emitted by a wastewater treatment plant. *Environmental Science and Pollution Research*, **25**, 9254–9264. https://doi.org/doi.org/10.1007/s11356-017-0024-0

Cooper E., Siewicki T. and Phillips, K. (2008). Preliminary risk assessment database and risk ranking of pharmaceuticals in the environment. *Science of the Total Environment*, **398**(1–3), 26–33. https://doi.org/10.1016/j.scitotenv.2008.02.061

Daughton, C. G. and Ruhoy, I. S. (2008). The afterlife of drugs and the role of PharmEcovigilance. *Drug Safety*, **31**, 1069–1082. https://doi.org/10.2165/0002018-200831120-00004

Daughton C. G. and Ruhoy I. S. (2009). Environmental footprint of pharmaceuticals: the significance of factors beyond direct excretion to sewers. *Environmental Toxicology and Chemistry*, **28**(12), 2495–2521, https://doi.org/10.1897/08-382.1

Dobrzynska J., Dabrowska M., Olchowski R., Zieba E. and Dobrowolski R. (2021). Development of a method for removal of platinum from hospital wastewater by novel ion-imprinted mesoporous organosilica. *Journal of Environmental Chemical Engineering*, **9**(4), 105302, https://doi.org/10.1016/j.jece.2021.105302

Donovan E., Unice K., Roberts J., Harris M. and Finiley, B. (2008). Risk of gastrointestinal disease associated with exposure to pathogens in the water of the lower Passaic river. *Applied and Environmental Microbiology*, **74**(4). https://doi.org/10.1128/AEM.00601-07

Ebrahim M., Gravel D., Thabet C., Abdesselam K., Paramalingam S. and Hyson C. (2016). Antimicrobial use and antimicrobial resistance trends in Canada: 2014. *Canada Communicable Disease Report*, **42**(11), 227–231, https://doi.org/10.14745/ccdr.v42i11a02

Edelsberg J., Weycker D., Barron R., Li X., Wu H., Oster G., Badre S., Langeberg W. J. and Weber D. J. (2014). Prevalence of antibiotic resistance in US hospitals. *Diagnostic Microbiology and Infectious Disease*, **78**(3), 255–262, https://doi.org/10.1016/j.diagmicrobio.2013.11.011

Efaq A. N., Tahman, N., Nagao, H., Al-Gheethi, A. A., Shahadat, M., & Kadir, M. O. A. (2015). Supercritical carbon dioxide as non-thermal alternative technology for safe handling of clinical wastes. *Environmental Processes*, **2**, 797–822. https://doi.org/doi.org/10.1007/s40710-015-0116-0

Efaq, A. N., Rahman, N., Nagao, H., Al-Gheethi, A. and Kadir, M. O. A. (2017). Inactivation of Aspergillus spores in clinical wastes by supercritical carbon dioxide. *Arabian Journal for Science and Engineering*, **42**, 39–51. https://doi.org/doi.org/10.1007/s13369-016-2087-5

Ekhaise F. and Omavwoya B. (2008). Influence of hospital wastewater discharged from University of Benin Teaching Hospital (UBTH), Benin City on its receiving environment. *American-Eurasian Journal Agriculture & Environmentl Science* **4**(4), 484–488.

Emmanuel E., Perrodin Y., Keck G., Blanchard J.-M. and Vermande P. (2002). Effects of Hospital Wastewater on Aquatic Ecosystem. XXVIII Congreso Interamericano de Ingeniería Sanitaria y Ambiental Cancún, 27 Al 31 de Octubre, October 7, 2002, México.

Escher B. I., Baumgartner R., Koller M., Treyer K., Lienert J. and McArdell C. S. (2011). Environmental toxicology and risk assessment of pharmaceuticals from hospital wastewater. *Water Research*, **45**(1), 75–92, https://doi.org/10.1016/j.watres.2010.08.019

Fedaku S., Merid Y., Beyene H., Teshome W. and Gebre-Selassie S. (2015). Assessment of antibiotic- and disinfectant-resistant bacteria in hospital wastewater, south Ethiopia: a cross-sectional study. *The Journal of Infection in Developing Countries*, **9**(2), 149–156. https://doi.org/10.3855/jidc.4808

Focazio M. J., Kolpin D. W., Barnes K. K., Furlong E. T., Meyer M. T., Zaugg S. D., Barber L. B. and Thurman M. E. (2008). A national reconnaissance for pharmaceuticals and other organic wastewater contaminants in the United States–II) Untreated drinking water sources. *Science of the Total Environment*, **402**(2–3), 201–216.

Galvin S., Boyle F., Hickey O., Vellinga A., Morris D. and Cormican, M. (2010). Enumeration and characterization of antimicrobial-resistant Escherichia coli bacteria in effluent from municipal, hospital, and secondary treatment facility sources. *Applied and Environmental Microbiology*, **76**(14), 4772–4779.

González Peña O. I., López Zavala M. Á. and Cabral Ruelas H. (2021). Pharmaceuticals market, consumption trends and disease incidence are not driving the pharmaceutical research on water and wastewater. *International Journal of Environmental Research and Public Health*, **18**(5), 1–37, https://doi.org/10.3390/ijerph18052532

Hall M. J., Schwartzman A., Zhang J. and Liu X. (2017). Ambulatory surgery centers: United States 2010. *National Health Statistics Reports, Centers for Disease Control & Prevention*, **102**, 1–15.

Harris S., Morris C., Morris D., Cormican M. and Cummins E. (2014). Antimicrobial resistant Escherichia coli in the municipal wastewater system: effect of hospital effluent and environmental fate. *Science of the Total Environment*, **468–469**, 1078–1085, https://doi.org/10.1016/j.scitotenv.2013.09.017

Hofer, U. (2019). The cost of antimicrobial resistance. Nature Reviews Microbiology, 17(January), 1–3. https://doi.org/https://doi.org/10.1038/s41579-018-0125-x

Ibrahim C., Hassen A., Pothier P., Mejri S. and Hammami, S. (2018). Molecular detection and genotypic characterization of enteric adenoviruses in a hospital wastewatr. *Environmental Science and Pollution Research*, **25**, 10977–10987. https://doi.org/doi.org/10.1007/s11356-018-1399-2

Iwanowicz L., Blazer V., Guy C., Pinkney A., Mullican J., and Alvarez, D. (2009). Reproductive health of bass in the Potomac, USA, drainage: Part 1. Exploring the effects of proximity to wastewater treatment plant discharge. *Environmental Toxicology and Chemistry*, **28**(5), 1072–1083. https://doi.org/10.1897/08-433.1

Iweriebor B., Iwu C., Obi L., Uchechukwu N. and Okoh, A. (2015). Multiple antibiotic resistances among Shiga toxin producing Escherichia coli O157 in.feces of dairy cattle farms in Eastern Cape of South Africa. BMC Microbiology. https://doi.org/10.1186/s12866-015-0553-y

Iweriebor B. C., Gaqavu S., Obi L. C., Nwodo U. U. and Okoh A. I. (2015). Antibiotic susceptibilities of Enterococcus species isolated from hospital and domestic wastewater effluents in Alice, Eastern Cape Province of South Africa. *International Journal of Environmental Research and Public Health*, **12**(4), 4231–4246.

Jeffries K., Nelson E., Jackson L., and Habibi, H. (2009). Basin-wide impacts of compounds with estrogen-like activity on longnose dace (Rhinichthys cataractae) in two prairie rivers of Alberta, Canada. *Environmental Toxicology and Chemistry*, **27**(10), 2042–2052. https:// doi.org/10.1897/07-529.1

Kannan K., Yun S. H., Ostaszewski A., McCabe J. M., Mackenzie-Taylor D. and Taylor A. B. (2008). Dioxin-like toxicity in the Saginaw River watershed: polychlorinated dibenzo-p-dioxins, dibenzofurans, and biphenyls in sediments and floodplain soils from the Saginaw and Shiawassee Rivers and Saginaw Bay, Michigan, USA. *Archives of Environmental Contamination and Toxicology*, **54**(1), 9–19.

Kaur, R., Yadav, B. and Tyagi, R. D. (2020). Microbiology of hospital wastewater. In Current Developments in Biotechnoogy and Bioengineering: Environmental and Health Impact of Hospital Wastewater, pp. 103–148. https://doi.org/doi.org/10.1016/B978-0-12-819722-6.00004-3

Katouli M., Thompson J., Gündoğdu A., & Stratton, H. M. (2012). Antibiotic resistant bacteria in hospital wastewaters and sewage treatment plants. Science Forum and Stakeholder Engagement: Building Linkages, Collaboration and Science Quality, 225–229.

Khan U. and Nicell J. (2015). Human health relevance of pharmaceutically active compounds in drinking water. *The APPS Journal*, **17**, 558–585. https://doi.org/doi.org/10.1208/s12248-015-9729-5

Kleywegt S., Pileggi V., Lam Y. M., Elises A., Puddicomb A., Purba G., Di Caro J. and Fletcher T. (2016). The contribution of pharmaceutically active compounds from healthcare facilities to a receiving sewage treatment plant in Canada. *Environmental Toxicology and Chemistry*, **35**(4), 850–862, https://doi.org/10.1002/etc.3124

Kraupner N., Hutinel M., Schumacher K., Gray D. A., Genheden M., Fick J., Flach C. F. and Larsson D. G. J. (2021). Evidence for selection of multi-resistant E. coli by hospital effluent. *Environment International*, **150**, 106436, https://doi.org/10.1016/j.envint.2021.106436

Kümmerer K. (2001). Drugs in the environment: emission of drugs, diagnostic aids and disinfectants into wastewater by hospitals in relation to other sources – a review. *Chemosphere*, **45**(6–7), 957–969, https://doi.org/10.1016/S0045-6535(01)00144-8

Korzeniewska E. and Harnisz M. (2013). Extended-spectrum beta-lactamase (ESBL)-positive Enterobacteriaceae in municipal sewage and their emission to the environment. *Environmental Management*, **128**, 904–911. https://doi.org/doi.org/10.1016/j.jenvman.2013.06.051

Kostich M. S. and Lazorchak J. M. (2008). Risks to aquatic organisms posed by human pharmaceutical use. *Science of the Total Environment*, **389**(2-3), 329–33.

Kostich M., Batt A., and Lazorchak, J. (2014). Concentrations of prioritized pharmaceuticals in effluents from 50 large wastewater treatment plants in the US and implications for risk estimation. *Environmental Pollution*, **184**, 354–359. https://doi.org/ 10.1016/j.envpol.2013.09.013

Lamba M. and Ahammad S. (2017). Sewage treatment effluents in Delhi: A key contributor of β-lactam resistant bacteria and genes to the environment. *Chemosphere*, **188**, 249–256. https://doi.org/doi.org/10.1016/j.chemosphere.2017.08.133

Le Corre K. S., Ort C., Kateley D., Allen B., Escher B. I. and Keller J. (2012). Consumption-based approach for assessing the contribution of hospitals towards the load of pharmaceutical residues in municipal wastewater. *Environment International*, **45**(1), 99–111, https://doi.org/10.1016/j.envint.2012.03.008

Lenz K., Hann S., Koellensperger G., Stefanka Z., Stingeder G., Weissenbacher N., Mahnik S. N. and Fuerhacker M. (2005). Presence of cancerostatic platinum compounds in hospital wastewater and possible elimination by adsorption to activated sludge. *Science of the Total Environment*, **345**(1–3), 141–152, https://doi.org/10.1016/j.scitotenv.2004.11.007

Li D., Qi R., Yang M., Zhang Y. and Yu, T. (2011). Bacterial community characteristics under long-term antibiotic selection pressures. *Water Research*, **45**(18), 6063–6073. https://doi.org/10.1016/j.watres.2011.09.002

Lien L., Lan P., Chuc N., HOA N., Nhung P., Thoa N., Diwan V., Tamhankar A., and Stålsby, L. (2017). Antibiotic resistance and antibiotic resistance genes in Escherichia coli isolates from hospital wastewater in Vietnam. *International Journal of Enviromental Research and Public Health*, **14**(7), 699. https://doi.org/doi.org/10.3390/ijerph14070699

Malakar A., Snow D. D. and Ray C. (2019). Irrigation water quality–A contemporary perspective. *Water*, **11**(7), 1482. https://doi.org/10.3390/w11071482

Mandal S., Ghosh A. and Pati B. (2015). Dissemination of antibiotic resistance in methicillin-resistant Staphylococcus aureus and vancomycin-resistant S aureus strains isolated from hospital effluents. *American Journal of Infection Control*, **43**(12), e87–e88. https://doi.org/doi.org/10.1016/j.ajic.2015.08.015

Mi R., Patidar R., Farenhorst A., Cai Z., Sepehri S., Khafipour E. and Kumar A. (2019). Detection of fecal bacteria and antibiotic resistance genes in drinking water collected from three first nations communities in manitoba, Canada. *FEMS Microbiology Letters*, **366**(6), 1–9, https://doi.org/10.1093/femsle/fnz067

Neudorf K. D., Huang Y. N., Ragush C. M., Yost C. K., Jamieson R. C. and Truelstrup Hansen L. (2017). Antibiotic resistance genes in municipal wastewater treatment systems and receiving waters in Arctic Canada. *Science of the Total Environment*, **598**, 1085–1094, https://doi.org/10.1016/j.scitotenv.2017.04.151

Obasi A., Amaeze N., and Osoko, D. (2014). Microbiological and toxicologiacl assessment of pharmaceutical wastewater from the Lagos Megacity, Nigeria. Chinese Journal of Biology. https://doi.org/10.1155/2014/63814

Okojokwu O. H. and Inabo S. (2014). Parasitological profile of raw wastewater and the efficacy of biosandfilter in reduction of parasite ova and cysts. *Journal of Applied Science Environment Management*, **18**(1), 59.

Oliveira T. S., Murphy M., Mendola N., Wong V., Carlson D. and Waring L. (2015). Characterization of pharmaceuticals and personal care products in hospital effluent and waste water influent/effluent by direct-injection LC-MS-MS. *Science of the Total Environment*, **518–519**, 459–478, https://doi.org/10.1016/j.scitotenv.2015.02.104

Ort C., Lawrence M. G., Reungoat J., Eaglesham G., Carter S. and Keller J. (2010). Determining the fraction of pharmaceutical residues in wastewater originating from a hospital. *Water Research*, **44**(2), 605–615, https://doi.org/10.1016/j.watres.2009.08.002

Picão R., Cardoso J., Campana E., Nicoletti A., Petrolini F., Assis D., Juliano L. and Gales, A. (2013). The route of antimicrobial resistance from the hospital effluent to the environment: focus on the occurrence of KPC-producing Aeromonas spp. and Enterobacteriaceae in sewage. *Diagnostic Microbiology and Infectious Disease*, **76**(1), 80–85. https:// doi.org/10.1016/j.diagmicrobio.2013.02.001

Prado T., Silva D., Guilayn W., Rose T., Gaspar A. M. and Miagostovich, M. (2011). Quantification and molecular characterization of enteric viruses detected in effluents from two hospital wastewater treatment plants. *Water Research*, **45**, 1287–1297. https://doi.org/doi.org/10.1016/j.watres.2010.10.012

Robinson I., Junqua G., Van Coillie R. and Thomas O. (2007). Trends in the detection of pharmaceutical products, and their impact and mitigation in water and wastewater in North America. *Analytical and Bioanalytical Chemistry*, **387**(4), 1143–1151, https://doi.org/10.1007/s00216-006-0951-y

Rodriguez-Mozaz S., Lucas D. and Barceló D. (2017). Full-scale plants for dedicated treatment of hospital effluents. In: Hospital Wastewaters. The Handbook of Environmental Chemistry, P. Verlicchi (Ed.), Springer, Dordrecht. https://doi.org/doi.org/10.1007/698_2017_13

Santoro D., Cardoso A., Coutinho F., Pinto L., Vierira R., Albano R. and Clementino, M. (2015). Diversity and antibiotic resistance profiles of Pseudomonads from a hospital wastewater treatment plant. *Journal of Applied Microbiology*, **119**(6), 1527–1540. https:// doi.org/10.1111/jam.12936

Singh R. D., Jurel S. K., Tripathi S., Agrawal K. K. and Kumari R. (2014). Mercury and other biomedical waste management practices among dental practitioners in India. *BioMed Research International*, **2014**, https://doi.org/10.1155/2014/272750

Srinivasan A., LOpez-RIbot J. & Ramasubramanian A. (2014). Overcoing antifungal resistance. *Drug Discovery Today: Technologies*, **11**, 65–71. https://doi.org/doi.org/10.1016/j.ddtec.2014.02.005

Stewart J. R., Fleming L. E., Fleisher J. M., Abdelzaher A. M. and Lyons M. M. (2011). Waterborne pathogens. Marine Pollution and Human Health, Issues in Environmental Science and Technology. Royal Society of Chemistry, London, 25–67.

Szekeres E., Baricz A., Chiriac C., Farkas A., Opris O., Soran M., Andrei A., Rudi K., Balcazar J., Dragos N. and Corman, C. (2017). Abundance of antibiotics, antibiotic resistance genes and bacterial community composition in wastewater effluents from different Romanian hospitals. *Environmental Pollution*, **225**, 304–315. https://doi.org/doi.org/10.1016/j.envpol.2017.01.054

Tchbanoglous G., Burton F. and Stensel H. (2003). Wastewater Engineering: Treatment and Reuse. Mc-Graw Hill Science.

Timraz K., Xiong Y., Al Qarni, H. and Hong, P. (2017). Removal of bacterial cells, antibiotic resistance genes and integrase genes by on-site hospital wastewater treatment plants: surveillance of treated hospital effluent quality. *Environmental Science: Water Resources*, **3**, 293–303. https://doi.org/10.1039/C6EW00322B

Thorpe K., Benstead R., Eccles P., Maack G., William T. and Tyler, C. (2008). A practicable laboratory flow-through exposure system for assessing the health effects of effluents in fish. *Aquatic Toxicology*, **88**(3), 164–172. https://doi.org/10.1016/j.aquatox.2008.04.005

Tiwari B., Sellamuthu B., Ouarda Y., Drogui P., Tyagi R. and Buelna, G. (2017). Review on fate and mechanism of removal of pharmaceutical pollutants from wastewater using biological approach. *Bioresource Technology*, **224**, 1–12. https://doi.org/doi.org/10.1016/j.biortech.2016.11.042

ul Haq M. R., Metcalfe C., Li H. and Parker W. (2012). Discharge of pharmaceuticals into municipal sewers from hospitals and long-term care facilities. *Water Quality Research Journal of Canada*, **47**(2), 140–152.

Varela A. R., Ferro G., Vredenburg J., Yanik M., Vieira L., Rizzo L., Lameiras C. and Manaia C. M. (2013). Vancomycin resistant enterococci: from the hospital effluent to the urban wastewater treatment plant. *Science of the Total Environment*, **450–451**, 155–161, https://doi.org/10.1016/j.scitotenv.2013.02.015

Verburg I., van Veelen H. P. J., Waar K., Rossen J. W. A., Friedrich A. W., Leal L. H., García-Cobos S. and Schmitt H. (2021). Effects of clinical wastewater on the bacterial community structure from sewage to the environment. *Microorganisms*, **9**(4), 1–16, https://doi.org/10.3390/microorganisms9040718

Verlicchi P. (ed.) (2018). Hospital Wastewaters. Vol. **60**. Springer. http://link.springer.com/10.1007/978-3-319-62178-4

Verlicchi P., Galletti A., Petrovic M., Petrovic M., Barceló D. and Barceló, D. (2010). Hospital effluents as a source of emerging pollutants: An overview of micropollutants and sustainable treatment options. *Journal of Hydrology*, **389**, 416–428. https://doi.org/10.1016/J.JHYDROL.2010.06.005

Verlicchi P., Al Aukidy M., Galletti A., Petrovic M. and Barceló D. (2012). Hospital effluent: investigation of the concentrations and distribution of pharmaceuticals and environmental risk assessment. *Science of the Total Environment*, **430**, 109–118, https://doi.org/10.1016/j.scitotenv.2012.04.055/ (accessed 2021.09.10)

Verlicchi P., Galletti A. and Al Aukidy M. (2013). Hospital wastewaters: quali-quantitative characterization and strategies for their treatment and disposal. In: Wastewater Management and Reuse, Sharma S. and Sanghi R. (eds.), Springer, Dordrecht, pp. 225–251, https://doi.org/10.1007/978-94-007-4942-9_8

Watkinson A., Murby E., Kolpin D. and Constanzo S. (2009). The occurrence of antibiotics in an urban watershed: from wastewater to drinking water. *Science of the Total Environment*, **407**(8), 2711–2723, https://doi.org/10.1016/j.scitotenv.2008.11.059

Wiest L., Chonova T., Bergé A., Baudot R., Bessueille-Barbier F., Ayouni-Derouiche L. and Vulliet E. (2018). Two-year survey of specific hospital wastewater treatment and its impact on pharmaceutical discharges. *Environmental Science and Pollution Research*, **25**(10), 9207–9218, https://doi.org/10.1007/s11356-017-9662-5

Zhang Q. H., Yang W. N., Ngo H. H., Guo W. S., Jin P. K., Dzakpasu M., Yang S. J., Wang Q., Wang X. C. and Ao D. (2016). Current status of urban wastewater treatment plants in China. *Environment International*, **92–93**, 11–22, https://doi.org/10.1016/j.envint.2016.03.024

doi: 10.2166/9781789062625_0101

Chapter 6

Groundwater contamination by hospital effluent with a focus on carbamazepine and sulfamethoxazole

*María Alejandra Villa Arroyave[1], Mariana Gómez Ospina[1], María Alejandra Cruz Bolaños[1], Thiago Victor Medeiros do Nascimento[1] and Jan Willem Foppen[1,2]**

[1]Water Science and Engineering Department, IHE-Delft Institute for Water Education, Delft, Netherlands
[2]Water Resource Section, Department of Civil Engineering and Geoscience, Delft University of Technology, Delft, Netherlands
**Corresponding author: j.foppen@un-ihe.org*

ABSTRACT

Groundwater can be polluted by a variety of anthropogenic points or line sources of contaminants. One such source is hospital wastewater, as this can carry a wide range of pharmaceuticals active compounds (PhACs). Due to their low removal efficiency, PhACs are frequently detected in the effluent of conventional wastewater treatment plants (WWTPs). Owing to their limited natural attenuation in soils, these compounds have the capacity to reach the underlying aquifer system and remain there for long periods. In this chapter, we reviewed the occurrence and processes related to two of the most frequently detected PhACs in surface and subsurface waters worldwide, namely sulfamethoxazole (SMX) and carbamazepine (CBZ), which are being studied for inclusion in the so-called European Union review of Annex I and II of the Groundwater Directive (GWD). Among the results found, sorption of SMX and CBZ in groundwater depended on organic matter content, while the degradability of both compounds was enhanced under reductive conditions. Maximum concentrations of SMX and CBZ in groundwater were 1600 ng/L and 2325 ng/l, respectively. Both contaminants were mainly reported in sandy/gravel aquifers influenced by WWTP effluents. A particular concern for human health arose since many of these aquifers were also used for drinking water supply. Additionally, not only the parental compounds but also their metabolites imposed environmental risks to groundwater bodies. The persistent and wide distribution of these substances and their metabolites in the environment highlighted the importance of regulatory steps.

Keywords: Carbamazepine, groundwater, hospital wastewater, pollution, sulfamethoxazole

CONTENTS

Editor(s): Nadeem A Khan, Viola Vambol, Sergij Vambol, Nastaran Mozaffari and Niloofar Mozaffari

6.1 INTRODUCTION

General awareness of hospital effluents is rising due to the significant risk of contamination they impose on the ecosystem and water resources. Hospital wastewater (HWW) might take several paths depending on local disposal management (Figure 6.1).

More often, HWW is discharged into the public sewage system and then treated together with municipal wastewater. Although a combined treatment is commonly adopted, the average concentration of several pollutants in HWW is approximately 150 times higher than in municipal wastewater (Cidlinová *et al.*, 2018). Most pollutants experience high removal rates in wastewater treatment plants (WWTPs). However, some organic compounds may pass through the process either unaffected or metabolised and/or transformed, which could even reduce the overall treatment efficiency (García Galán *et al.*, 2012; Lapworth *et al.*, 2018; Wang *et al.*, 2019). HWW is considered a significant producer of emerging organic compounds (EOCs), a term used to describe not yet regulated contaminants under current or future concern due to new toxicological data (Stuart *et al.*, 2012). Among these compounds, the occurrence in the environment of many pharmaceuticals used in human medical care has been emphasised due to their intrinsic biological activity and their potential deleterious effect on human health (Lesser *et al.*, 2018). The pharmaceutically active compounds (PhACs) comprise the pharmaceuticals and their transformation products, the pharmaceutically active metabolites (Heberer, 2002).

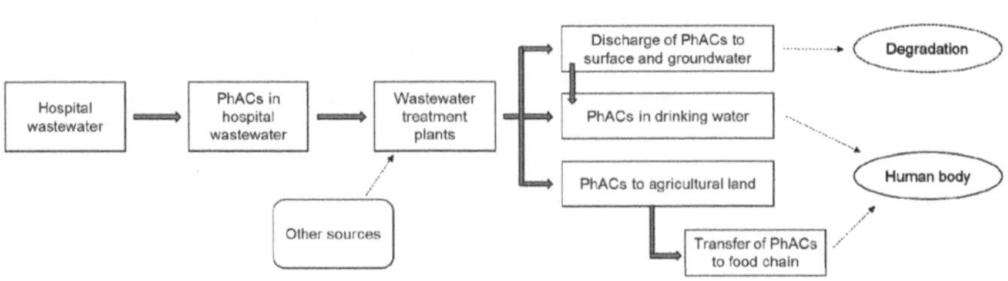

Figure 6.1 Main pathway of PhACs originated from HWW (adapted from Khan *et al.* 2021).

Box 6.1: Important terms explained

Sorption: Mechanism for attenuation of contaminants in the subsurface. Sorption of hydrophobic organic chemicals is an instantaneous and equilibrium process, giving rise to both sorption and desorption (Sposito, 2008). Usually, a sorption isotherm is experimentally determined to define the relationship between the concentration of a contaminant in the sorbed and aqueous phase. The linear isotherm is the simplest way to describe sorption through the sorption coefficient (K_d). However, some organic compounds exhibit non-linear sorption, better described by the Freundlich isotherm (defined by K_f and n). These parameters determine the sorption capacity and intensity, respectively (Appelo & Postma, 2005; Sposito, 2008).

Decay: Decay or degradation implies the change of the molecule's chemical structure (Appelo & Postma, 2005) and, eventually, it results in the formation of one or more metabolites. Decay rates of PhACs are primarily due to microbial degradation, as some studies have demonstrated by comparing half-lives of pharmaceuticals under sterile and non-sterile soils (Kodešová *et al.*, 2016, 2020). Due to the importance of biodegradation, microbial activity and abundance play a key role. The microbial conditions are controlled by soil properties, structure, water content, aeration, and nutrient content. Seasonal variations in soil properties and different redox potentials due to the presence of nitrate, sulfate, or humic acids can influence xenobiotic biodegradation as well (Appelo & Postma, 2005).

Half-life: The time required for a particular compound to decay to half of its initial concentration is the half-life ($t_{1/2}$) (Sposito, 2008). In first-order kinetic reactions, the half-life is independent of the initial concentrations and can be estimated as $\ln(2)/k$. The coefficient k is the rate constant or specific rate (Appelo & Postma, 2005; Sposito, 2008).

Once in the environment, the transport and fate of contaminants in porous media is mainly controlled by decay and sorption (Biel-Maeso *et al.*, 2019; McCance *et al.*, 2018) (for an explanation, see Box 6.1). In general, groundwater samples have about two orders of magnitude lower PhAC concentrations compared to WWTP effluent (Kibuye *et al.*, 2019), which enforces the notion that these pollutants are indeed subject to removal (Bexfield *et al.*, 2019; D'Alessio *et al.*, 2018; Hebig *et al.*, 2017). However, the assumption that pharmaceuticals would be effectively removed from the soil by sorption in the unsaturated zone or by biodegradation has been proven wrong (Rozman *et al.*, 2017). These beliefs are still considered because of the simplified approach of the degradation processes, where the fate of metabolites and reversibility in the reaction are generally not considered (Rodríguez-Escales & Sanchez-Vila, 2016).

Groundwater Directive (GWD) 2006/118/EC, included within the Water Framework Directive (WFD) 2000/60, was established to ensure the good status of groundwater bodies by preventing or limiting groundwater pollution. Consequently, the identification and monitoring of currently unregulated contaminants with the potential to hinder the achievement of environmental objectives were deemed necessary (Kozel *et al.*, 2019; Lapworth *et al.*, 2019). Therefore, the voluntary Groundwater Watch List (GWWL) was created and conceived as a dynamic process. First, prioritised compounds in the list should be studied to collect sufficient monitoring data and later evaluated on whether a substance should be included or not in Annexes I or II of the GWD via the List Facilitating process (Bunting *et al.*, 2021).

Currently, eight compounds are included in the watch list of substances for Union-wide monitoring (Article 8b of Directive 2008/105/EC); these are pharmaceuticals, including hormones (E1, E2, and EE2), antibiotics (ciprofloxacin and amoxicillin), and macrolides (roxithromycin, clarithromycin, and azithromycin).

This chapter is a compilation of the most recent and relevant work on the occurrence, transport, and fate of sulfamethoxazole (SMX) and carbamazepine (CBZ), two of the most frequently detected PhACs related to HWW, in subsurface waters worldwide. Owing to their limited natural attenuation in soils, these compounds can reach the underlying aquifer systems and be detected in surface water and groundwater (Bunting *et al.*, 2021; Elliott *et al.*, 2018; Heberer, 2002; Jurado *et al.*, 2020; Zainab *et al.*, 2020). SMX and CBZ were selected, as they are in the List Facilitating Annex I/II review of the GWD. This list was created to identify emerging pollutants for which groundwater quality standards or threshold values should be set. Hence, for pollutants shown to pose a risk, a mechanism exists for establishing threshold values (Annex II of Directive 2006/118/EC as amended by Directive 2014/80/EU) or groundwater quality standards (Annex I). Even though SMX and CBZ have been widely detected in groundwater for drinking water supply throughout the United States (Bexfield *et al.*, 2019), there is no official regulation to control the occurrence of these pharmaceuticals in the environment. The United States Environmental Protection Agency (EPA), in the framework of the Safe Drinking Water Act (SDWA), has a drinking water Contaminant Candidate List (CCL) developed through a three-phase Regulatory Determination Assessment process. However, the current CCL, composed of 97 chemicals and 12 microbial contaminants, excludes SMX and CBZ.

6.2 METHODOLOGY

We present an overview of data published in the past 20 years on SMX and CBZ. Literature searches were carried out from June to October 2021 in the Scopus and ScienceDirect databases for articles published in English. Research papers and literature reviews examining the occurrence, fate, and removal of studied pharmaceutical residues in groundwater were targeted. The search terms used were: [Hospital Effluents OR Hospital Wastewater] AND groundwater], [sulfamethoxazole AND groundwater AND hospitals], and [carbamazepine AND groundwater AND hospitals]. Furthermore, to ensure relevance, we only considered those papers with a citation score of 100 or above for publications older than 10 years.

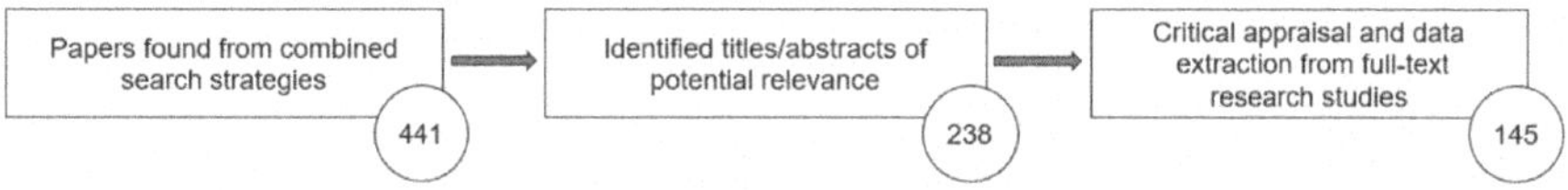

Figure 6.2 Selection of studies for inclusion in the overview.

All potentially relevant papers were obtained in full and critically appraised. They were filtered according to the information they could provide about the general chemical properties and behaviour of SMZ and CBZ, that is, their sorption, degradation pathways, metabolites formation, toxicity, and their occurrence in groundwater and aquatic environments. Literature on pharmaceutical pollution of veterinary residues or manure was excluded. As an additional strategy, references in collected articles were scanned to identify other potentially relevant papers. Finally, 145 publications were selected (Figure 6.2), including reviews, laboratory experiments, and a range of environmental studies on surface water, groundwater, and drinking water.

First, we present a general overview of the studied compounds, followed by a conceptual description of the parameters that influence their occurrence and fate in groundwater. Next, the associated ecotoxicity and potential risks to human health, described by the literature, are summarised. Finally, based on the analysed data, we discuss the future outlook for SMX and CBZ in groundwater to identify key implications, existing gaps, and suggestions for steps to reduce SMX and CBZ loads in the aquatic environment.

6.3 RESULTS

6.3.1 Sulfamethoxazole (SMX)

SMX is one of the most commonly used antimicrobial drugs for human and veterinary medicine and the most widely detected antibiotic in soils and groundwater (Bexfield *et al.*, 2019; Chen *et al.*, 2017; Rodríguez-Escales & Sanchez-Vila, 2016). SMX mainly enters wastewater via human excretion as its human transformation product N-acetyl-SMX and/or N-SMX-glucuronide, as only 5–13% of the absorbed dose leaves the body in an unaltered form (Ven *et al.*, 1995). Thus, both are readily cleaved back to SMX while passing through a sewer system or during wastewater treatment (Göbel *et al.*, 2005; Müller *et al.*, 2012). In HWW, SMX concentrations of 390 ng/L have been reported, with removal rates of 22–33% (Brown *et al.*, 2006).

The SMX molecule (Table 6.1) is polar, hydrophilic ($S_w = 610$ mg/L) and highly mobile ($K_{ow} = 0.89$), weakly acidic ($pK_a = 1.6$ or 5.7), with amphoteric properties and a low soil sorption tendency (Candela *et al.*, 2016; Rodríguez-Escales & Sanchez-Vila, 2016). Under neutral conditions, SMX is negatively charged (Candela *et al.*, 2016). SMX's hydrophilic behaviour, together with a low log K_{ow} (<1) and its sorption coefficient, suggests that it weakly sorbs, is highly mobile, and is easily leached into groundwater (Chen *et al.*, 2017; FAO, 2000). Many studies have confirmed the fast migration of SMX from soil to groundwater bodies (Strauss *et al.*, 2011) and found SMX may migrate relatively

Table 6.1 Physicochemical properties of SMX, CBZ, and their human metabolites, described in the chapter.

Pharmaceutical	CAS	Formula	Water Solubility (S_w)[a]	Log K_{ow}[a]	pK_a[a]	Molecular Structure[b]
Sulfamethoxazole (SMX)	723–46–6	$C_{10}H_{11}N_3O_3S$	610 mg/L	0.89	1.60 5.70	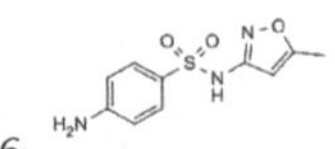
Carbamazepine (CBZ)	298–46–4	$C_{15}H_{12}N_2O$	18 mg/L	2.45	13.9–15.96	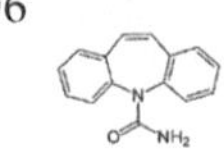

[a]*United States National Center for Biotechnology Information* (https://pubchem.ncbi.nlm.nih.gov/).
[b]*Adapted from König et al. (2016) and Rodríguez-Escales and Sanchez-Vila (2016).*

deep. In saturated porous media, it can be transported over considerable distances up to depths of 1300 m (Archundia *et al.*, 2017).

SMX linear sorption coefficients (K_d; see Box 6.1) ranged from 0.6 to 7.4 L/kg. In particular, K_d values for clayey soils have been found mainly below 2 L/kg (Archundia *et al.*, 2019; D'Alessio *et al.*, 2018; Srinivasan & Sarmah, 2014). Nevertheless, SMX often exhibits a non-linear sorption isotherm, better fitted by the Freundlich model, with Freundlich coefficients (K_f; see Box 6.1) around 0.634 L/kg (Zuo *et al.*, 2021). K_d values were negatively correlated with soil pH (Avisar *et al.*, 2009), which means that slight differences in adsorption could be expected between soils with different pH values. For instance, at pH values higher or equal to 7.2, SMX existed mainly in its anionic form and, consequently, was more soluble in aqueous solutions (Biel-Maeso *et al.*, 2019; Jurado *et al.*, 2020). Under these conditions, SMX lost its cationic charge, being less adsorbed onto the soil. However, SMX hydrophobic partition became minor when background ions were present, enhancing its absorption (Chen *et al.*, 2017). In addition, SMX sorption was directly related to the organic matter content of the soil (Doretto *et al.*, 2014) and, thus, would be less adsorbed in sandy soils than in clay soils (Biel-Maeso *et al.*, 2019).

As SMX sorption onto aquifer sediments could usually be neglected, the observed reduction in SMX concentration within the environment was mainly associated with biodegradation processes. Koba *et al.* (2016) and Liu *et al.* (2010) confirmed this by observing a faster degradation in non-sterile soils compared to sterile soils. Literature on SMX degradation rate reported half-lives from 9 to 18 days in field and laboratory-scale experiments (Gottschall *et al.*, 2012; Lin & Gan, 2011), all within the 'readily degradable' classification of the FAO (2000). Baumgarten *et al.* (2011), Biel-Maeso *et al.* (2019) and Kodešová *et al.* (2020) found that due to biodegradation, aerobic environments could result in shorter half-lives of SMX than anoxic or anaerobic environments. Consequently, SMX was more persistent in the absence of oxygen, although the shortest degradation time of SMX has been found in the presence of iron-reducing conditions (Rodríguez-Escales & Sanchez-Vila, 2016). In this sense, the degradation of SMX appeared to be redox-dependent (Banzhaf *et al.*, 2012; Baumgarten *et al.*,

2011; Heberer *et al.*, 2008; Hebig *et al.*, 2017) and mainly related to residence time rather than SMX persistence (Baumgarten *et al.*, 2011).

Furthermore, laboratory experiments showed shorter half-lives of SMX in topsoil than in substrate samples, as well as lower values of half-lives in cambisols and coarser substrates, compared to loess and marlite soils. These results could be related to microbial conditions in the different soils (Kodešová *et al.*, 2016). On top of this, readily degradable carbon sources may have enhanced biodegradation, as they could provide energy for heterotrophic biomass growth and metabolic activity (Müller *et al.*, 2012).

The transformation behaviour and the redox-sensitivity of SMX stipulate the importance of knowing redox and nitrate concentrations to understand the occurrence and transport of SMX in groundwater and the aquatic environment (Banzhaf *et al.*, 2012), as well as when studying SMX metabolites (Rodríguez-Escales & Sanchez-Vila, 2016). Under nitrate-reducing conditions, SMX degradation would result in two non-stable metabolites, 4-nitro-SMX and desamino-SMX, which were eventually retransformed to the parent compound once denitrification was absent (Rodríguez-Escales & Sanchez-Vila, 2016). Therefore, the presence of the former metabolites would be plausible in environments where nitrate and antibiotics leakage convergance was present (García Galán *et al.*, 2012). Rodríguez-Escales and Sanchez-Vila (2016) defined the SMX degradation process based on the reaction occurring by the re-oxidation of ferrous iron (II) to ferric iron (III) on the surface of goethite, which facilitated the break-up of the SMX molecule. This process resulted in the formation of four metabolites, controlled mainly by the redox state. Since the isoxazole ring was irreversibly broken, it would facilitate the eventual mineralisation of SMX.

Additionally, the low amount of organic carbon in the subsoil would limit the potential degradation pathways. Iron-reducing conditions would only be reached once the redox states ended. Upon reaching that state, the degradation of SMX was achieved within hours, being the shortest reported characteristic time for degradation of the SMX molecule. These metabolites presented the lowest risk for ecosystems (Rodríguez-Escales & Sanchez-Vila, 2016). Therefore, it was likely that the degradation efficiency of SMX in groundwater bodies could be attained by an engineered introduction of organic carbon in the aquifer that enhanced iron-reducing conditions (Rodríguez-Escales & Sanchez-Vila, 2016). The transformation process of SMX under reductive and aerobic conditions is summarised in Figure 6.3.

6.3.2 Carbamazepine (CBZ)

CBZ is a widely prescribed anticonvulsant drug used to treat epilepsy, trigeminal neuralgia, manic depression, and bipolar disorder (Bunting *et al.*, 2021; Koroša *et al.*, 2020). Approximately 72% of orally administered CBZ is absorbed and then is heavily metabolised by the liver; only about 1% of the absorbed dose leaves the body in an unaltered form (Zhang *et al.*, 2008). After being metabolised, CBZ is excreted through the urine and faeces (Jurado *et al.*, 2014) and introduced to the wastewater as unmodified CBZ or as its human transformation products, such as 10,11-dihydro-10,11-epoxy carbamazepine

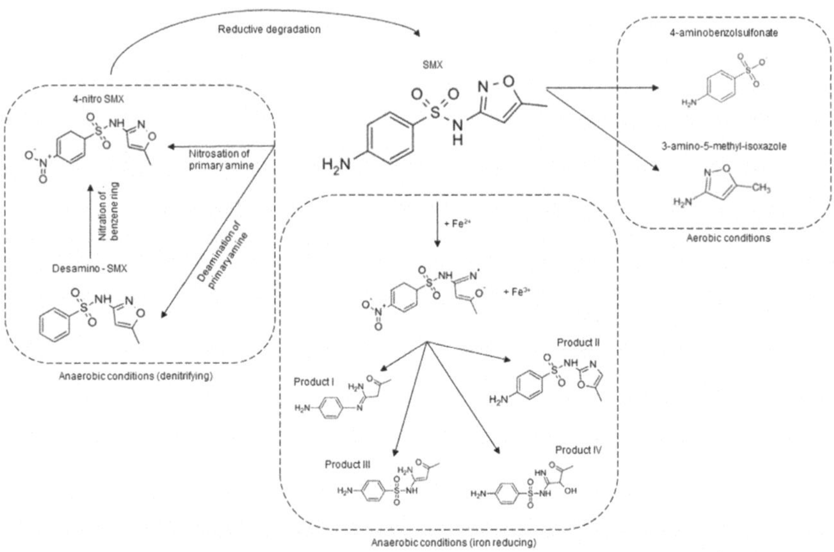

Figure 6.3 Reaction pathways of SMX under anaerobic (denitrification or iron-reducing) and aerobic conditions (adapted from Rodríguez-Escales and Sanchez-Vila, 2016).

(CBZ-EP), hydroxylated compounds 2-hydroxy carbamazepine (2OH-CBZ) and 3-hydroxy carbamazepine (3OH-CBZ), acridine (ACRIN) and acridone (ACRON) (Jurado *et al.*, 2014). Concentrations of CBZ in HWW ranged between 68 and 15 780 ng/L (Khan *et al.*, 2021; Park & Lee, 2018). The removal efficiency during processing in a conventional WWTP was frequently observed as less than 25% (Jelic *et al.*, 2012; Park & Lee, 2018; Zhang *et al.*, 2008).

The CBZ molecule (Table 6.1) is polar, uncharged, weakly acidic ($pK_a = 13.9$–15.96) and moderately hydrophobic ($S_w = 18$ mg/L). It has a characteristic stable heterocyclic structure that allows its persistence (Candela *et al.*, 2016; Koroša *et al.*, 2020). Since CBZ can be classified as a neutral substance, it is likely to interact with soil organic matter ($\log K_{ow} = 2.45$). Therefore, in the presence of soil with rich organic matter, its retardation increases, being considered as a slow-mobile compound (Chefetz *et al.*, 2008; Silver *et al.*, 2018).

As a consequence, CBZ could potentially leach into the aquifer when treated, or untreated wastewater is introduced into the environment (Yu *et al.*, 2013), and its persistence could be enhanced by its low K_f values and poor biodegradability (Biel-Maeso *et al.*, 2019). Furthermore, CBZ was detected in the vadose zone up to 24.2 m depth (Zentner *et al.*, 2015). Hence, this compound was frequently considered an anthropogenic marker in groundwater and the aquatic environment (Bizi, 2019) and was used in several studies to quantify wastewater leakage reaching groundwater and bank filtration (Dvory *et al.*, 2018; Hebig *et al.*, 2017; Koroša *et al.*, 2020; McCance *et al.*, 2018; Wolf *et al.*, 2012).

The concentration trends found by Silver *et al.* (2018) suggested that the primary attenuation mechanism of CBZ was sorption. Although there was a great variety of sorption estimations for CBZ, overall, it was shown that the sorption isotherms of CBZ were less linear in soils with high soil organic matter (Chefetz *et al.*, 2008; Kiecak *et al.*, 2020). A Freundlich isotherm model better described the non-linear sorption behaviour of CBZ (see Box 6.1). Biel-Maeso *et al.* (2019) found a relatively low sorption capacity for high-clay content soils (K_f=78 L/kg) decreasing further for sandy soils (K_f=39 L/kg). Weaker sorption was found by Chefetz *et al.* (2008), who reported K_f values of 11.46 to 12.63 L/kg for high organic matter soils. The different results suggested a higher dependency of CBZ sorption on soil type, with a positive correlation with organic carbon content (Hebig *et al.*, 2017; Kodešová *et al.*, 2020). In addition, Kiecak *et al.* (2019) and Schaffer *et al.* (2012) showed a negative correlation between retardation by sorption and pH due to the decrease of hydrophobic interactions of CBZ molecules with the sediments. Burke *et al.* (2014) focused on temperature but did not find any temperature dependency attenuation for CBZ.

Regarding its degradation, field and laboratory-scale degradation rate studies found CBZ to be a recalcitrant compound (Jelic *et al.*, 2012; Park & Lee, 2018; Silver *et al.*, 2018). The absence of electron-donating groups in CBZ could cause its resistance to degradation (Nakada *et al.*, 2007; Yu *et al.*, 2006). Many researchers did not record noticeable degradation of CBZ during the length of their studies (Kodešová *et al.*, 2016). They indicated that CBZ could remain intact after 8–10 years of travel through the subsurface under anoxic saturated or aerobic unsaturated flow conditions (Drewes *et al.*, 2002). The reported half-life values for CBZ ranged from 35 to 120 days, in aerobic conditions (Henzler *et al.*, 2014; Li *et al.*, 2013), to 355 to 1624 days in non-sterile soil column experiments (Dalkmann *et al.*, 2014). Following FAO (2000) classification for half-lives and the degree of degradability of a compound in soils, CBZ was fairly to very slightly degradable.

As for SMX, nitrate concentration, redox conditions and residence time in the aquifer influenced the fate of CBZ and the formation of transformational products in the environment (Jurado *et al.*, 2014). Banzhaf *et al.* (2012) found CBZ degraded when nitrate levels were high in the sediments. Wiese *et al.* (2011) reported CBZ removal of 50% under iron-reducing conditions with 54 days of mean flow time, while under oxic conditions, the efficiency of removal was around 16% with 23 days of mean flow time.

The metabolisation process under anaerobic conditions resulted in two independent reduction pathways. When CBZ reduction occurred abiotically, more than 85% of CBZ was transformed to 2H-CBZ, and the remaining 15% was metabolised to 16H-CBZ. Therefore, the most dominant and stable reduction metabolite was 2H-CBZ, which can undergo biotic transformation to $C_{16}H_{19}N_2$ (König *et al.*, 2016). Whereas the degradation intermediates that emerged from aerobic conditions included 10,11-dihydro-10-hydroxycarbamazepine, carbamazepine-10,11-epoxide (transient intermediate), acridine, and two more metabolites with deduced molecular structures from their fragmentation patterns. One of these transformation products, acridone-N-carbaldehyde, was

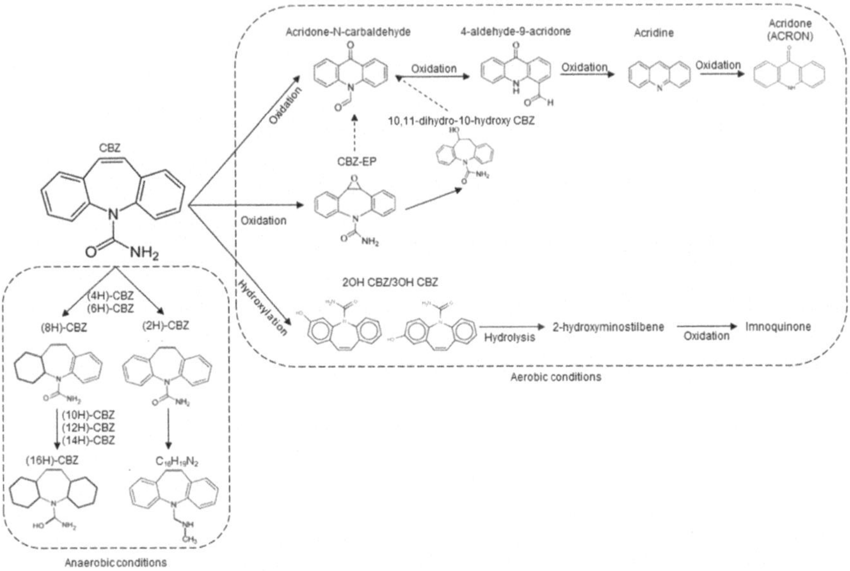

Figure 6.4 Reaction pathways of CBZ under anaerobic and aerobic conditions (adapted from König *et al.*, 2016; Li *et al.*, 2013).

the most abundant intermediate of microbial transformation and exhibited recalcitrant behaviour. The last metabolite, 4-aldehyde-9-acridone, was characteristically more polar than the rest (Li *et al.*, 2013). The degradation intermediates of CBZ and their reaction pathways under reductive and aerobic conditions are summarised in Figure 6.4.

6.3.3 Occurrence of SMX and CBZ

As previously mentioned, there are multiple and complex pathways followed by PhACs in the environment (Arnold *et al.*, 2014). Similar to most pharmaceuticals, CBZ and SMX occurrence was mainly attributed to hospital, household, and pharmaceutical industry effluents (Baena-Nogueras *et al.*, 2017; Vambol *et al.*, 2021). The spatial distribution in the occurrence of these contaminants was linked to their availability and consumption trends. For instance, in Europe alone, CBZ was prescribed more than 6 million times in 2018 (Zorginstituut Nederland, 2018). However, this information was not readily available in several developing countries (Fekadu *et al.*, 2019).

The main route of these compounds was through sewage treatment plants and sludge disposal. Upon release from WWTPs, CBZ and SMX were widely distributed in surface water and infiltrated into groundwater (Arnold *et al.*, 2014; Bunting *et al.*, 2021; Cardoso-Vera *et al.*, 2021). A major contributor to these compounds entering into groundwater was the reuse of untreated/ treated sewage in agriculture, landscape irrigation, and artificial groundwater recharge (Jurado *et al.*, 2014; Lesser *et al.*, 2018; Lopez *et al.*, 2015). Once in the aquatic environment, pollutants reached deeper soil layers or groundwater through leaching (Babić & Mutavdžić Pavlović, 2013), and the physicochemical properties of the contaminants determined their fate (Cardoso-Vera *et al.*, 2021; Zuo *et al.*, 2021).

The aquifer system in the Mezquital Valley in Mexico, composed of volcanic and sedimentary materials, was the most prominent example of the deep connection between pharmaceuticals' occurrence and wastewater reuse. There, 70% of the wastewater generated by the urban area was recycled to provide crop nutrients, causing high recharge into the aquifers. SMX and CBZ, together with other commonly prescribed PhACs, were frequently detected in springs and groundwater, affecting nearby groundwater abstractions for human consumption (Lesser *et al.*, 2018). Furthermore, Lopez *et al.* (2015) found that the occurrence of emerging contaminants, such as CBZ, was greatest in groundwater samples near urban activities associated with agriculture. These activities were mainly located on top of sedimentary formations and alluvial aquifers.

Additionally, CMZ and SMX occurred during artificial recharge for drinking water supply, which took place in a sand-gravel fluvial terrace (Hrkal *et al.*, 2018). River delta aquifers, associated with heavily polluted rivers, network sewage loss, and water supply loss, as main recharge sources, evidenced the presence of CBZ (Jurado *et al.*, 2014).

Table 6.2 summarises the average, minimum and maximum concentrations of SMX and CBZ in groundwater exposed to contamination mainly by WWTP effluents (more details in Table 6.4). For example, Bradley *et al.* (2014) reported a maximum concentration of 1600 ng/L for SMX, found in a shallow unconfined aquifer system influenced by WWTP effluent. Similarly, a maximum concentration of CBZ (2325 ng/L) was recorded in an aquifer/aquitard system affected by treated sewage water (Müller *et al.*, 2012).

On the other hand, the minimum concentration value over the detection limit of SMX was found in a partly confined aquifer system used for drinking water supply (Zuo *et al.*, 2021). For CBZ, the minimum value was found in an aquifer system potentially affected by the wastewater effluent from a psychiatric hospital (Xiang *et al.*, 2018). Therefore, these concentrations might be explained by a combination of the physical properties of the aquifer systems and exposure time.

Table 6.2 Concentrations reported in groundwater for carbamazepine and sulfamethoxazole. A detailed list of concentrations per published study can be found in Table 6.4 of this chapter.

Compound	**Concentrations (ng/L)**[a]		
	Mean	**Minimum**	**Maximum**
Sulfamethoxazole (SMX)	23.7	0.2	1600
Carbamazepine (CBZ)	60.4	0.3	2325

[a]Concentration values were computed as the average of the mean, minimum of the minimum, and maximum of the maximum values reported.

Table 6.3 Some selected examples of adverse effects of SMX and CBZ on non-target organisms.

Pharmaceutical	Non-target Species	Effects	Reference
SMX	The marine bacterium (*Vibrio fischeri*), a freshwater invertebrate (*Daphnia magna*), and the medaka fish (*Oryzias latipes*)	Acute adverse effects on the species with high hazard coefficients	Kim *et al.* (2007)
	Juvenile nile tilapia (*Oreochromis niloticus*)	Alters the liver function and induces inflammatory response	Hu *et al.* (2021)
	Embryos of zebrafish (*Damio rerio*)	Affects growth development, induces malformation of the embryo, triggers an inflammatory response, and reduces immune defence in healthy young fish	Liu *et al.* (2020); Yan *et al.* (2018)
	Green algae (*Chlorella vulgaris*)	High toxicity effects in the algae, with increased toxicity towards the increase in salinity levels in the environment	Borecka *et al.* (2016)
CBZ	The marine bacterium (*Vibrio fischeri*), a freshwater invertebrate (*Daphnia magna*), and the medaka fish (*Oryzias latipes*)	Moderate adverse effects on the species	Kim *et al.* (2007)
	Crustacean (*Daphnia similis)*	Inhibits moulting, delays reproduction, and reduces fecundity	Chen *et al.* (2019a)
	Chinese mitten crab (*Eriocheir sinensis*)	Disrupts moulting hormone signalling and inhibits growth at environmentally relevant concentrations	Chen *et al.* (2019b)
	Embryos of zebrafish (*Damio rerio*)	Inhibits hatching and produces growth retardation effects	van den Brandhof and Montforts (2010)
	Embryos of rats (*Rattus norvegicus*)	Embryo development retardation and damage	Piersma *et al.* (1998)
	Juvenile rainbow trout (*Oncorhynchus mykiss)*	Alters body development of the juvenile fish	Li *et al.* (2011)

Table 6.4 Mean, minimum and maximum concentration of SMX and CBZ reported in groundwater per published study.

Compound	Reference	Concentration (ng/L)		
		Minimum	Maximum	Mean
Sulfamethoxazole (SMX)	Biel-Maeso *et al.* (2019)	47.8	251.5	33.8
	Candela *et al.* (2016)	4.0	18.0	9.0
	Elliott *et al.* (2018)	7.0	965.0	129.0
	Qin *et al.* (2020)	0.7	20.1	7.0
	Zainab *et al.* (2021)		29.1	
	Oberleitner *et al.* (2020)		50.0	
	Boy-Roura *et al.* (2018)		28.6	6.1
	Loos *et al.* (2010)		38.0	2.0
	Ma *et al.* (2015)		250.0	15.0
	Lapworth *et al.* (2018)	0.7	34.0	
	van Stempvoort *et al.* (2013)		458.0	
	Bartelt-Hunt *et al.* (2011)		20.0	
	McEachran *et al.* (2016)		21.0	
	Vulliet and Cren-Olivé (2011)		53.9	3.0
	Cabeza *et al.* (2012)	23.2	83.9	
	Teijon *et al.* (2010)	2.0	117.0	45.6
	Jiang *et al.* (2014)		105.7	26.1
	Yang *et al.* (2018)	7.9	25.7	
	Zuo *et al.* (2021)	0.2	0.2	
	Loos *et al.* (2010)		38.0	2.0
	Bradley *et al.* (2014)		1600.0	
	Jurado *et al.* (2020)			5.9
	Average	10.4	200.4	23.7
Carbamazepine (CBZ)	Biel-Maeso *et al.* (2019)		162.0	8.5
	Fram and Belitz (2011)		40.0	420.0
	Loos *et al.* (2010)		390.0	12.0
	Sacher *et al.* (2001)	2.0	900.0	
	Wolf *et al.* (2012)		35.0	
	Vulliet and Cren-Olivé (2011)		10.4	
	Lopez *et al.* (2015)	10.0	100.0	
	López-Serna *et al.* (2013)	7.0	136.0	115.0
	Cabeza *et al.* (2011)		62.4	
	Petrović *et al.* (2014)			3.4
	Zemann *et al.* (2014)		500.0	74.0
	Xiang *et al.* (2018)	0.3	7.1	2.6
	Stuart *et al.* (2014)		40.0	

(*continued*)

Table 6.4 *(Continued)*

Compound	Reference	Concentration (ng/L)		
		Minimum	Maximum	Mean
	Rozman *et al.* (2017)		753.0	
	Reh *et al.* (2013)	2.4	346.7	38.4
	K'oreje *et al.* (2020)	0.5	100.0	
	Ma *et al.* (2018)		4.4	0.7
	Ma *et al.* (2018)		12.5	1.1
	Lin *et al.* (2015)	0.4	37.9	
	Lesser *et al.* (2018)		370.0	
	Lapworth *et al.* (2015)	10.0	30.0	
	Koroša *et al.* (2016)	5.0	88.0	17.4
	Jurado *et al.* (2014)		17.8	7.1
	Jurado *et al.* (2014)		136.0	115.4
	Elliott *et al.* (2018)	9.9	1000.0	38.4
	van Driezum *et al.* (2019)	1.8	27.9	
	Park and Lee (2018)	52.3	91.0	67.2
	Candela *et al.* (2016)	3.0	134.0	46.0
	Bunting *et al.* (2021)		500.0	
	Branchet *et al.* (2019)	5.0	335.0	
	Bexfield *et al.* (2019)		162.0	
	Müller *et al.* (2012)		2325.0	
	Average	7.8	285.6	60.4

Overall, concentrations of SMX and CBZ were high. In a pan-European reconnaissance of organic contaminants performed by Loos *et al.* (2010), SMX and CBZ were detected in 24.4% and 42.1% of 164 groundwater samples, respectively. Similar results were found by Bexfield *et al.* (2019), who reported SMX in 23% of 47 groundwater samples and Hillebrand *et al.* (2012), who reported CBZ in 57.3% of 157 spring water samples.

6.3.4 Associated risks

6.3.4.1 Ecotoxicity

In general, the reported concentrations of SMX and CBZ in groundwater were very low compared to relevant acute toxicology endpoints, such as EC50 or LC50 (Lapworth *et al.*, 2012). Acute median effective/lethal concentrations (E/LC50) were usually higher for SMX than CBZ (Kar & Roy, 2010; Kim *et al.*, 2007) and, thus, higher concentrations of SMX were needed to have adverse/lethal effects on 50% of the test population within a specifically designated period. However, chronic exposure to these compounds might have affected groundwater-dependent ecosystems since aquifers could be considered cumulative receptors of organic pollutants, and pharmaceuticals were biologically active substances (Berninger & Brooks, 2010). Chronic toxicity effects were usually shown as subtle changes in the physiology, behaviour, or biochemistry of non-targeted organisms (Peake *et al.*, 2016). Some examples of the ecotoxicological effects of SMX and CBZ are provided in Table 6.3.

Moreover, the environmental risk analysis of SMX and CBZ should include the effects of transformational products, as they were heavily metabolised within the body after consumption, around 70% and 90%, respectively (Jurado *et al.*, 2014; Ven *et al.*, 1995). During transformation, the toxicophore from the parent compound may remain intact, or, on rare occasions, the metabolite may have a new toxicophore with higher or similar toxicity (de Jongh *et al.*, 2012). Therefore, when limited information about the ecotoxicity of metabolites is available, as a worst-case scenario, their toxicity should be assumed equal to that of their parent compound (Peake *et al.*, 2016). In addition, the potential bioaccumulation and the impact of synergistic effects by mixtures of compounds must also be considered to assess the real environmental risk (Bunting *et al.*, 2021).

Specifically on sulfonamides, Carvalho and Santos (2016) highlighted the growing interest in this group of antibiotics in the environment because low levels could enhance the development of antibiotic-resistant bacteria and genes. The development of bacterial resistance is a growing global challenge and presents a public health concern (Cidlinová *et al.*, 2018; Kovalakova *et al.*, 2020; Lapworth *et al.*, 2018). SMX-resistant bacteria could impact the soil and subsurface biodiversity, and affect the natural processes of pollution attenuation, that is, causing changes in denitrification, methanogenesis, sulfate reduction, nutrient cycles, and organic matter degradation (Bunting *et al.*, 2021; Kovalakova *et al.*, 2020; Lapworth *et al.*, 2018). It could also trigger endocrine disruption (Park & Lee, 2018), acute poisoning or gene mutation in some non-targeted organisms, especially when mixed with other antibiotics (Chen *et al.*, 2017). SMX was classified by Kovalakova *et al.* (2020) as very toxic to photosynthetic organisms, with aquatic plants and cyanobacteria being the most sensitive classes.

In the case of the neuroactive drug CBZ, the most frequently reported effects included oxidative stress and protein damage (Li *et al.*, 2011), endocrine disruption in *Daphnia magna*, as in *Daphnia similis* (Chen *et al.*, 2019b), growth retardation, and heart abnormalities in rats (Piersma *et al.*, 1998) and fish embryos (van den Brandhof & Montforts, 2010). Most of these effects may be caused by the inhibition of foraging capacity and food assimilation, which changed the energy budget of the affected organism (Lamichhane *et al.*, 2013). Furthermore, one of its primary metabolites, CBZ-EP, with similar antiepileptic properties as its parent compound, caused a similar spectrum of foetal malformations, like soft tissue defects and skeletal defects in rats (Piersma *et al.*, 1998). Several researchers have suggested that the teratogenic effects of CBZ exposure were due to its metabolites rather than the parent compound (Li *et al.*, 2013).

There are multiple methodologies to assess the general environmental risk, one of them was proposed by the European Medicines Agency (EMA). The ecotoxicological risks could be expressed as a risk quotient (RQ) between the predicted environmental concentration (PEC) (substance's exposure level) or measured environmental concentration (MEC), and the ecological safety threshold at which no pharmacological effect was expected to occur for a specific organism, referred to as predicted no-effect concentration (PNEC).

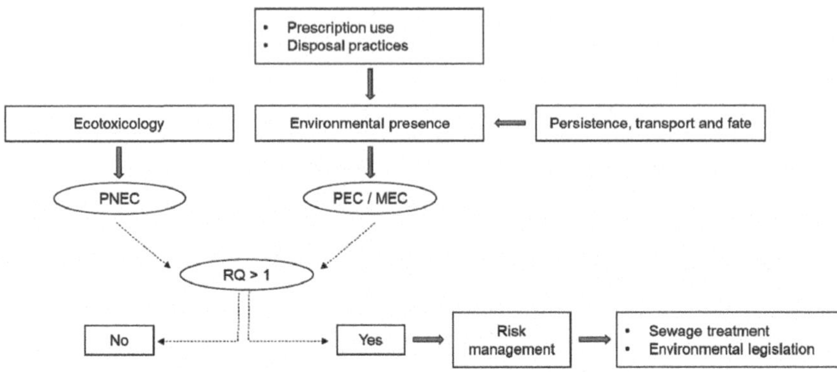

Figure 6.5 Aspects of EMA risk assessment in the environment (adapted from Peake *et al.*, 2016).

However, this methodology was biased by the available toxicity information on the pharmaceutical (von der Ohe *et al.*, 2011), and it used two assumptions; one was that the most sensitive species defined the ecosystem sensitivity, and the second was that protecting ecosystem structure protected community function (Joint Research Centre, 2003).

The PEC could be defined by considering annual pharmaceutical consumption, the removal rate of the WWTP, population, and dilution factor, or as the 95th percentile of the maximum MEC (von der Ohe *et al.*, 2011). The values of PNECs could be found in peer review literature and, if not available, they should be obtained by dividing the lowest value of the relevant available toxicity data (EC50 or LC50) by an arbitrarily defined constant, usually 1000 (Joint Research Centre, 2003; Ma *et al.*, 2015). In the literature, SMX values of PNECs for aquatic environments have been reported as 146 ng/L (Kim *et al.*, 2007), 27 ng/L (Ma *et al.*, 2015), and 52 ng/L (von der Ohe *et al.*, 2011). CBZ PNECs were frequently higher, between 31 600 ng/L (Kim *et al.*, 2007) and 32 000 ng/L (von der Ohe *et al.*, 2011).

The resulting RQs were a useful measure to characterise the potential ecological risk of a stressor (Kim *et al.*, 2007) and could be classified into insignificant (<0.1), low (0.1 to 1.0), moderate (>1.0 to 10), and high (>10). Nevertheless, in general, if the RQs reach values equal to or higher than 1, the substances should be considered to have the potential to cause adverse ecological effects (Commission, 2003; see Figure 6.5).

The RQ for SMX in groundwater was reported by Ma *et al.* (2015) as 9 (moderate risk), but Jurado *et al.* (2020) found that, individually, all SMX and studied metabolites would be far below 1 (insignificant to low risk). Hoffmann-La Roche company determined an RQ of 0.387 (low risk) for chronic antibiotic exposure during their Environmental Risk Assessment (Straub, 2016). Kim *et al.* (2007) calculated the RQ for SMX, based on a conservative computation of PEC, to be 6.3 (moderate risk), and for CBZ to be 0.0044 (insignificant). Ferrari *et al.* (2003) found PEC/PNEC ratios for CBZ of 29.2 and 47 (high risk), assuming PECs of WWTP effluents. Some of these studies indicated that SMX and CBZ might pose a moderate or high environmental risk, depending on the used PEC/MEC value and the local conditions of occurrence.

6.3.4.2 Drinking water

Groundwater is the world's main freshwater reservoir, providing at least 50% of drinking water and 43% of irrigation water, while it supports the base flow of rivers, lakes, and wetlands. Thus, the occurrence and fate of SMX and CBZ in groundwater are of particular concern to human health. If present in surface extractions or groundwater supply wells, the traces of the pharmaceuticals could enter the human body through direct consumption of drinking water or indirectly with food. Several studies confirmed the presence of SMX and CBZ in drinking water wells, for example, Fram and Belitz (2011) and Hrkal *et al.* (2018), while many others observed these compounds in water bodies affected by the reuse of wastewater in agricultural fields, for example, Lesser *et al.* (2018) and Lopez *et al.* (2015). Hellauer *et al.* (2018) reported CBZ in the artificially recharged groundwater, which provides 70% of the drinking water of Berlin, although observed concentrations of SMX and CBZ could not cause effects on humans or achieve a pharmacological effect. For instance, the therapeutic dose for CBZ is about 800 to 1200 mg/day, and assuming the highest concentration of CBZ (2325 ng/l reported in 2012 by Müller *et al.* 2012), to produce a pharmacological effect, it would be necessary to drink 344.1 to 516.1 m^3 of contaminated water if removal of CBZ during drinking water treatment was omitted. In the case of SMX, Hoffmann-La Roche defined an acceptable daily exposure of 10 mg/person/day by all routes of intake (Straub, 2016). Considering the highest concentration (1600 ng/L reported by Bradley *et al.* 2014) and a default drinking water consumption of 2 L/day, without removal of the pollutant, the daily intake of SMX would amount to 3.2×10^{-3} mg/person. Besides, SMX and CBZ had a large buffer between the maximum concentrations detected and the human health benchmarks of 100 000 ng/L and 40 000 ng/L, respectively (Bexfield *et al.*, 2019).

Moreover, de Jesus Gaffney *et al.* (2015) performed a human health risk characterisation for pharmaceuticals based on RQ methodology by substituting the PNEC value with the respective Drinking Water Equivalent Level (DWEL). The DWEL can be defined by the toxicology of the compound and parameters related to the gastrointestinal absorption rate, frequency of absorption, and consumer body weight. The analysis done with Lisbon's drinking water supply samples took into account maximum detected concentrations to provide a conservative 'worst-case' scenario approach. All RQ values for SMZ and CBZ, estimated for different life stages, were found to be below 0.1 (insignificant risk). Jurado *et al.* (2020) determined the RQ for SMZ and the group of sulfonamide antibiotics to be below 0.1 (insignificant risk) with a similar methodology. Nonetheless, this did not exclude the possibility of other risks due to chronic exposure to SMZ and CBZ through drinking water (le Corre *et al.*, 2012).

Studies have attempted to identify the risk to human health due to continuous exposure to organic compounds, such as the studied PhACs, in drinking water from surface and groundwater sources. However, there is limited toxicological information available. The synergistic effects of the presence of multiple pharmaceuticals and metabolites are yet to be further studied and understood (Pereira *et al.*, 2015), and this is likely to impact the overall risk assessment results. One of the main and current lines of research is the potential impact of antibiotic-resistant bacteria on human health since they could lead to the transfer of resistance factors to human pathogens (le Corre *et al.*, 2012), and their occurrence is one of the most critical public health concerns of this century, according to the World Health Organization (Kovalakova *et al.*, 2020).

6.4 DISCUSSION AND OUTLOOK

6.4.1 Treatments for removing SMX and CBZ

Overall, we found that the low removal efficiency of SMX and CBZ at conventional WWTPs was connected to their occurrence in groundwater due to the disposal of sewage water and sludge. The differences in technology and management at treatment plants might have influenced their abundance (Cardoso-Vera *et al.*, 2021). Indeed, the remediation of contaminated groundwater is not a simple task, enforced by the long residence time of some compounds in the aquifer (Mutiyar & Mittal, 2014; Zainab *et al.*, 2021). The precautionary principle calls for minimising contaminant loads in the environment and drinking water (Aus Der Beek & Bergmann, 2016). Therefore, it is clear that current treatment techniques, including conventional mechanical-biological plants and managed aquifer recharge (MAR) systems, should be upgraded.

Treatments such as ultrafiltration, reverse osmosis, and UV disinfection could not remove SMX and CBZ (Candela *et al.*, 2016). Both compounds also persisted during MAR implementation, such as riverbank filtration and sequential MAR (Hamann *et al.*, 2016; van Driezum *et al.*, 2019). Among other alternatives, soil aquifer treatment (SAT) had the potential to control some pharmaceuticals with higher removal efficiencies (61.8% for SMX), although, for CBZ, an increase in the concentration was observed instead of removal (He *et al.*, 2016).

Nonetheless, D'Alessio *et al.* (2018) showed a removal capacity for CBZ greater than 98% and up to 75% for SMX using a tertiary treatment (filtration followed by disinfection). Similarly, some biological (treatment/ozonation, membrane bioreactor/reverse osmosis and membrane bioreactor/granular activated carbon) and chemical methods (photocatalytic degradation and advanced oxidation/biodegradation) removed more than 98% of CBZ, according to Cardoso-Vera *et al.* (2021) and Kozub *et al.* (2021). Riverbank filtration sites under reducing conditions registered the removal of 51% of CBZ concentrations due to the increase of travel time (König *et al.*, 2016; Oberleitner *et al.*, 2020), while other studies reported that CBZ was susceptible to degradation by photolysis and could be eliminated by more than 90% by aerobic biofilter; such as the surface layer of a recharge zone (Williams & McLain, 2012).

Besides the efforts to remove SMX and CBZ at WWTPs, depending on the physicochemical state of the aquifer, these compounds might naturally attenuate, as described in previous sections. For example, concentration reductions of 87.5% (SMX) and 44% (CBZ) were found in groundwater samples influenced by WWTP effluents (Candela *et al.*, 2016). However, due to the variable hydrogeological conditions of the aquifers affected by wastewaters, natural attenuation could be significantly lower, especially in aquifers with low organic content. In addition, as stated by Wiese *et al.* (2011), the removal efficiency is field site-specific.

The development and application of removal techniques should take advantage of the physicochemical properties of the compound. For instance, SMX and CBZ degraded faster during iron-reducing conditions (Rodríguez-Escales & Sanchez-Vila, 2016; Wiese *et al.*, 2011). Therefore, an engineered introduction of organic carbon in the aquifer could enhance iron-reducing conditions and consequently increase the degradation of the compound (Rodríguez-Escales & Sanchez-Vila, 2016). Special care should be taken regarding degradation into non-stable metabolites or ones with toxicological effects. Furthermore, transformation products of CBZ may be removed differently from aquatic environments. For instance, Kosjek *et al.* (2009) found that ACRIN could be efficiently removed under anoxic (90%) and anaerobic (92%) conditions. Conversely, ACRON was better removed under aerobic conditions (40%, compared to 23% in anoxic conditions) (Jurado *et al.*, 2014).

Finally, the improvement of current techniques, or the development of new wastewater treatment options, have been primary solutions in order to prevent further contaminant accumulation in the environment (Sousa *et al.*, 2018). Advanced treatment steps, such as ozonation, activated carbon, advanced oxidation, reverse osmosis, and membrane filtration processes, were usually required to effectively eliminate more recalcitrant pharmaceuticals (including SMX and CBZ) from wastewater (Hübner *et al.*, 2014; Khan *et al.*, 2021).

6.4.2 European Union groundwater directive (GWD)

The selection of substances to be included, whether in the GWWL or the List Facilitating Annex I and II process, was based on the integrated water score (ranked list IV), which considered the groundwater leaching potential score and the ecotoxicological hazard. In 2019, the Voluntary Groundwater List was published along with a ranked list of substances where SMX and CBZ were positioned in the top 25 pharmaceuticals and PFAS (per- and poly-fluoroalkyl substances). However, after applying the selection criteria, SMX and CBZ were deselected from the Groundwater Watch List process based on sufficient monitoring data (Carvalho *et al.*, 2015) and entered the 'List Facilitating Annex I/II review'. SMX and CBZ are two of the most studied organic pollutants in groundwater, and sufficient information was collected throughout this review about their detection in groundwater, mobility, persistence, pathways into the environment, and toxicity (Kozel *et al.*, 2019).

Since both compounds (SMX and CBZ) entered the 'List Facilitating Annex I/II review', they should be evaluated by the European Commission for possible inclusion in Annex II of the GWD (Kozel *et al.*, 2019). If included, the Member States should establish threshold values for these pollutants under the GWD objectives and mandatory monitoring of pollution indicators. The determination of threshold values would have to be supported by a control mechanism for the data collected, analytical considerations, and natural or anthropogenic background levels, as established in the Directive 2006/118/EC.

Moreover, the transformation products of CBZ have been found relevant in ecotoxicology studies. To date, there are relatively few studies on the topic, and no formal regulations to control, monitor, or report these contaminants in the aquatic environment, nor drinking water, in the European Union jurisdiction (Bunting *et al.*, 2021). However, no information was found indicating interest to evaluate metabolites for inclusion in the prioritisation list (GWWL) or the GWD Annexes.

APPENDIX

See Table 6.4.

REFERENCES

Appelo C. A. J. and Postma D. (2005). Geochemistry, Groundwater and Pollution, 2nd edn, CRS Press, London.

Archundia D., Duwig C., Lehembre F., Chiron S., Morel M. C., Prado B., Bourdat-Deschamps M., Vince E., Aviles G. F. and Martins J. M. F. (2017). Antibiotic pollution in the Katari subcatchment of the Titicaca Lake: major transformation products and occurrence of resistance genes. *Science of the Total Environment*, **576**, 671–682. https://doi.org/10.1016/j.scitotenv.2016.10.129

Archundia D., Duwig C., Spadini L., Morel M. C., Prado B., Perez M. P., Orsag V. and Martins J. M. F. (2019). Assessment of the sulfamethoxazole mobility in natural soils and of the risk of contamination of water resources at the catchment scale. *Environment International*, **130**, 1–10. https://doi.org/10.1016/j.envint.2019.104905

Arnold K. E., Brown A. R., Brown A. R., Ankley G. T. and Sumpter J. P. (2014). Medicating the environment: assessing risks of pharmaceuticals to wildlife and ecosystems. *Philosophical Transactions of the Royal Society B: Biological Sciences*, **369**(1656), 1–11. https://doi.org/10.1098/rstb.2013.0569

Aus Der Beek T. and Bergmann A. (2016). Pharmaceuticals in the Environment: Global Occurrence and Potential Cooperative Action Under the Strategic Approach to International Chemicals Management (SAICM), German Environment Agency, Dessau-Roßlau, Germany. https://www.researchgate.net/publication/330934183

Avisar D., Lester Y. and Ronen D. (2009). Sulfamethoxazole contamination of a deep phreatic aquifer. *Science of the Total Environment*, **407**(14), 4278–4282. https://doi.org/10.1016/j.scitotenv.2009.03.032

Babić S. and Mutavdžić Pavlović D. (2013). Analysis of PhACs in solid environmental samples (soil, sediment, and sludge). In: Comprehensive Analytical Chemistry, Petrovic M., Perez S. and Barcelo, D. (eds), Elsevier B.V., Oxford, Vol. **62**, pp. 129–167, https://doi.org/10.1016/B978-0-444-62657-8.00005-7

Baena-Nogueras R. M., González-Mazo E. and Lara-Martín P. A. (2017). Degradation kinetics of pharmaceuticals and personal care products in surface waters: photolysis vs biodegradation. *Science of the Total Environment*, **590–591**, 643–654. https://doi.org/10.1016/j.scitotenv.2017.03.015

Banzhaf S., Nödler K., Licha T., Krein A. and Scheytt T. (2012). Redox-sensitivity and mobility of selected pharmaceutical compounds in a low flow column experiment.

Science of the Total Environment, **438**, 113–121. https://doi.org/10.1016/j.scitotenv.2012.08.041

Bartelt-Hunt S., Snow D. D., Damon-Powell T. and Miesbach D. (2011). Occurrence of steroid hormones and antibiotics in shallow groundwater impacted by livestock waste control facilities. *Journal of Contaminant Hydrology*, **123**(3–4), 94–103. https://doi.org/10.1016/j.jconhyd.2010.12.010

Baumgarten B., Jährig J., Reemtsma T. and Jekel M. (2011). Long term laboratory column experiments to simulate bank filtration: factors controlling removal of sulfamethoxazole. *Water Research*, **45**(1), 211–220. https://doi.org/10.1016/j.watres.2010.08.034

Berninger J. P. and Brooks B. W. (2010). Leveraging mammalian pharmaceutical toxicology and pharmacology data to predict chronic fish responses to pharmaceuticals. *Toxicology Letters*, **193**(1), 69–78. https://doi.org/10.1016/j.toxlet.2009.12.006

Bexfield L. M., Toccalino P. L., Belitz K., Foreman W. T. and Furlong E. T. (2019). Hormones and pharmaceuticals in groundwater used as a source of drinking water across the United States. *Environmental Science and Technology*, **53**(6), 2950–2960. https://doi.org/10.1021/acs.est.8b05592

Biel-Maeso M., González-González C., Lara-Martín P. A. and Corada-Fernández C. (2019). Sorption and degradation of contaminants of emerging concern in soils under aerobic and anaerobic conditions. *Science of the Total Environment*, **666**, 662–671. https://doi.org/10.1016/j.scitotenv.2019.02.279

Bizi M. (2019). Activated carbon and the principal mineral constituents of a natural soil in the presence of Carbamazepine. *Water (Switzerland)*, **11**(11), 1–17. https://doi.org/10.3390/w11112290

Borecka M., Białk-Bielińska A., Haliński Ł. P., Pazdro K., Stepnowski P. and Stolte S. (2016). The influence of salinity on the toxicity of selected sulfonamides and trimethoprim towards the green algae Chlorella vulgaris. *Journal of Hazardous Materials*, **308**, 179–186. https://doi.org/10.1016/j.jhazmat.2016.01.041

Boy-Roura M., Mas-Pla J., Petrovic M., Gros M., Soler D., Brusi D. and Menció A. (2018). Towards the understanding of antibiotic occurrence and transport in groundwater: findings from the Baix Fluvià alluvial aquifer (NE Catalonia, Spain). *Science of the Total Environment*, **612**, 1387–1406. https://doi.org/10.1016/j.scitotenv.2017.09.012

Bradley P. M., Barber L. B., Duris J. W., Foreman W. T., Furlong E. T., Hubbard L. E., Hutchinson K. J., Keefe S. H. and Kolpin D. W. (2014). Riverbank filtration potential of pharmaceuticals in a wastewater-impacted stream. *Environmental Pollution*, **193**, 173–180. https://doi.org/10.1016/j.envpol.2014.06.028

Branchet P., Ariza Castro N., Fenet H., Gomez E., Courant F., Sebag D., Gardon J., Jourdan C., Ngounou Ngatcha B., Kengne I., Cadot E. and Gonzalez C. (2019). Anthropic impacts on Sub-Saharan urban water resources through their pharmaceutical contamination (Yaoundé Center Region, Cameroon). *Science of the Total Environment*, **660**, 886–898. https://doi.org/10.1016/j.scitotenv.2018.12.256

Brown K. D., Kulis J., Thomson B., Chapman T. H. and Mawhinney D. B. (2006). Occurrence of antibiotics in hospital, residential, and dairy effluent, municipal wastewater, and the Rio Grande in New Mexico. *Science of the Total Environment*, **366**(2–3), 772–783. https://doi.org/10.1016/j.scitotenv.2005.10.007

Bunting S. Y., Lapworth D. J., Crane E. J., Grima-Olmedo J., Koroša A., Kuczyńska A., Mali N., Rosenqvist L., van Vliet M. E., Togola A. and Lopez B. (2021). Emerging organic compounds in European groundwater. *Environmental Pollution*, **269**, 1–13. https://doi.org/10.1016/j.envpol.2020.115945

Burke V., Greskowiak J., Asmuß T., Bremermann R., Taute T. and Massmann G. (2014). Temperature dependent redox zonation and attenuation of wastewater-derived organic micropollutants in the hyporheic zone. *Science of the Total Environment*, **482–483**(1), 53–61. https://doi.org/10.1016/j.scitotenv.2014.02.098

Cabeza Y., Candela L., Teijón G. and Tamoh K. (2011). Presence of pharmaceuticals, personal care products, and priority substances (2008/105EU) in groundwater (Llobregat delta, Barcelona, Spain). *Geophysical Research Abstracts*, **13**, 2011–3524.

Cabeza Y., Candela L., Ronen D. and Teijon G. (2012). Monitoring the occurrence of emerging contaminants in treated wastewater and groundwater between 2008 and 2010. The Baix Llobregat (Barcelona, Spain). *Journal of Hazardous Materials*, **239–240**, 32–39. https://doi.org/10.1016/j.jhazmat.2012.07.032

Candela L., Tamoh K., Vadillo I. and Valdes-Abellan J. (2016). Monitoring of selected pharmaceuticals over 3 years in a detrital aquifer during artificial groundwater recharge. *Environmental Earth Sciences*, **75**(3), 1–13. https://doi.org/10.1007/s12665-015-4956-8

Cardoso-Vera J. D., Elizalde-Velázquez G. A., Islas-Flores H., Mejía-García A., Ortega-Olvera J. M. and Gómez-Oliván L. M. (2021). A review of antiepileptic drugs: part 1 occurrence, fate in aquatic environments and removal during different treatment technologies. In: Science of the Total Environment, Elsevier B.V., **768**, 1–15. https://doi.org/10.1016/j.scitotenv.2021.145487

Carvalho I. T. and Santos L. (2016). Antibiotics in the aquatic environments: a review of the European scenario. *Environment International*, **94**, 736–757. https://doi.org/10.1016/j.envint.2016.06.025

Carvalho R. N., Ceriani L., Ippolito A. and Lettieri T. (2015). Development of the First Watch List Under the Environmental Quality Standards Directive 2015, Institute for Environment and Sustainability, European Commission, Luxembourg. https://doi.org/10.2788/101376

Chefetz B., Mualem T. and Ben-Ari J. (2008). Sorption and mobility of pharmaceutical compounds in soil irrigated with reclaimed wastewater. *Chemosphere*, **73**(8), 1335–1343. https://doi.org/10.1016/j.chemosphere.2008.06.070

Chen K. L., Liu L. C. and Chen W. R. (2017). Adsorption of sulfamethoxazole and sulfapyridine antibiotics in high organic content soils. *Environmental Pollution*, **231**, 1163–1171. https://doi.org/10.1016/j.envpol.2017.08.011

Chen H., Gu X., Zeng Q. and Mao Z. (2019a). Acute and chronic toxicity of carbamazepine on the release of chitobiase, molting, and reproduction in daphnia similis. *International Journal of Environmental Research and Public Health*, **16**(2), 1–12. https://doi.org/10.3390/ijerph16020209

Chen H., Gu X., Zeng Q., Mao Z., Liang X. and Martyniuk C. J. (2019b). Carbamazepine disrupts molting hormone signaling and inhibits molting and growth of Eriocheir sinensis at environmentally relevant concentrations. *Aquatic Toxicology*, **208**, 138–145. https://doi.org/10.1016/j.aquatox.2019.01.010

Cidlinová A., Wittlingerová Z., Zimová M., Chrobáková T. and Petruželková A. (2018). Ecotoxicity of wastewater from medical facilities: a review. In: Scientia Agriculturae Bohemica, De Gruyter Open Ltd., Vol. **49**, Issue 1, pp. 26–31, https://doi.org/10.2478/sab-2018-0005

D'Alessio M., Onanong S., Snow D. D. and Ray C. (2018). Occurrence and removal of pharmaceutical compounds and steroids at four wastewater treatment plants in Hawai'i and their environmental fate. *Science of the Total Environment*, **631–632**, 1360–1370. https://doi.org/10.1016/j.scitotenv.2018.03.100

Dalkmann P., Siebe C., Amelung W., Schloter M. and Siemens J. (2014). Does long-term irrigation with untreated wastewater accelerate the dissipation of pharmaceuticals in soil? *Environmental Science and Technology*, **48**(9), 4963–4970. https://doi.org/10.1021/es501180x

de Jesus Gaffney V., Almeida C. M. M., Rodrigues A., Ferreira E., Benoliel M. J. and Cardoso V. V. (2015). Occurrence of pharmaceuticals in a water supply system and related human health risk assessment. *Water Research*, **72**, 199–208. https://doi.org/10.1016/j.watres.2014.10.027

de Jongh C. M., Kooij P. J. F., de Voogt P. and ter Laak T. L. (2012). Screening and human health risk assessment of pharmaceuticals and their transformation products in Dutch surface waters and drinking water. *Science of the Total Environment*, **427–428**, 70–77. https://doi.org/10.1016/j.scitotenv.2012.04.010

Doretto K. M., Peruchi L. M. and Rath S. (2014). Sorption and desorption of sulfadimethoxine, sulfaquinoxaline and sulfamethazine antimicrobials in Brazilian soils. *Science of the Total Environment*, **476–477**, 406–414. https://doi.org/10.1016/j.scitotenv.2014.01.024

Drewes J. E., Heberer T., Rauch-Williams T., Reddersen K. and Rauch T. (2002). Fate of pharmaceuticals during groundwater recharge evaluation of analytical methods for EDCs and PPCPs via interlaboratory comparison view project fate of pharmaceuticals during ground water recharge. *Ground Water Monitoring & Remediation*, **23**, 64–72. https://doi.org/10.1111/j.1745-6592.2003.tb00684.x

Dvory N. Z., Livshitz Y., Kuznetsov M., Adar E., Gasser G., Pankratov I., Lev O. and Yakirevich A. (2018). Caffeine vs. carbamazepine as indicators of wastewater pollution in a karst aquifer. *Hydrology and Earth System Sciences*, **22**(12), 6371–6381. https://doi.org/10.5194/hess-22-6371-2018

Elliott S. M., Erickson M. L., Krall A. L. and Adams B. A. (2018). Concentrations of pharmaceuticals and other micropollutants in groundwater downgradient from large on-site wastewater discharges. *PLoS ONE*, **13**(11), 1–17. https://doi.org/10.1371/journal.pone.0206004

European Commission - Joint Research Centre. (2003). Technical Guidance Document (TGD) in support of Commission Directive 93/67/EEC on Risk Assessment for new notified substances, Commission Regulation (EC) No 1488/94 on Risk Assessment for existing substances and Directive 98/8/EC of the 44 European Parliament and the Council concerning the placing of biocidal products on the market. (Patent No. PART II).

FAO (2000). FAO Pesticide Disposal Series 8 – Assessing Soil Contamination. Food and Agriculture Organization of the United Nations, Rome.

Fekadu S., Alemayehu E., Dewil R. and van der Bruggen B. (2019). Pharmaceuticals in freshwater aquatic environments: a comparison of the African and European challenge. *Science of the Total Environment*, **654**, 324–337. https://doi.org/10.1016/j.scitotenv.2018.11.072

Ferrari B., Paxéus N., lo Giudice R., Pollio A. and Garric J. (2003). Ecotoxicological impact of pharmaceuticals found in treated wastewaters: study of carbamazepine, clofibric acid, and diclofenac. *Ecotoxicology and Environmental Safety*, **55**(3), 359–370. https://doi.org/10.1016/S0147-6513(02)00082-9

Fram M. S. and Belitz K. (2011). Occurrence and concentrations of pharmaceutical compounds in groundwater used for public drinking-water supply in California. *Science of the Total Environment*, **409**(18), 3409–3417. https://doi.org/10.1016/j.scitotenv.2011.05.053

García Galán M. J., Díaz-Cruz M. S. and Barceló D. (2012). Removal of sulfonamide antibiotics upon conventional activated sludge and advanced membrane bioreactor treatment. *Analytical and Bioanalytical Chemistry*, **404**(5), 1505–1515. https://doi.org/10.1007/s00216-012-6239-5

Göbel A., Thomsen A., McArdell C. S., Joss A. and Giger W. (2005). Occurrence and sorption behavior of sulfonamides, macrolides, and trimethoprim in activated sludge treatment. *Environmental Science and Technology*, **39**(11), 3981–3989. https://doi.org/10.1021/es048550a

Gottschall N., Topp E., Metcalfe C., Edwards M., Payne M., Kleywegt S., Russell P. and Lapen D. R. (2012). Pharmaceutical and personal care products in groundwater, subsurface drainage, soil, and wheat grain, following a high single application of municipal biosolids to a field. *Chemosphere*, **87**(2), 194–203. https://doi.org/10.1016/j.chemosphere.2011.12.018

Hamann E., Stuyfzand P. J., Greskowiak J., Timmer H. and Massmann G. (2016). The fate of organic micropollutants during long-term/long-distance river bank filtration. *Science of the Total Environment*, **545–546**, 629–640. https://doi.org/10.1016/j.scitotenv.2015.12.057

He K., Echigo S. and Itoh S. (2016). Effect of operating conditions in soil aquifer treatment on the removals of pharmaceuticals and personal care products. *Science of the Total Environment*, **565**, 672–681. https://doi.org/10.1016/j.scitotenv.2016.04.148

Heberer T. (2002). Occurrence, fate, and removal of pharmaceutical residues in the aquatic environment: a review of recent research data. *Toxicology Letters*, **131**, 5–17. https://doi.org/10.1016/S0378-4274(02)00041-3

Heberer T., Massmann G., Fanck B., Taute T. and Dünnbier U. (2008). Behaviour and redox sensitivity of antimicrobial residues during bank filtration. *Chemosphere*, **73**(4), 451–460. https://doi.org/10.1016/j.chemosphere.2008.06.056

Hebig K. H., Groza L. G., Sabourin M. J., Scheytt T. J. and Ptacek C. J. (2017). Transport behavior of the pharmaceutical compounds carbamazepine, sulfamethoxazole, gemfibrozil, ibuprofen, and naproxen, and the lifestyle drug caffeine, in saturated laboratory columns. *Science of the Total Environment*, **590–591**, 708–719. https://doi.org/10.1016/j.scitotenv.2017.03.031

Hellauer K., Karakurt S., Sperlich A., Burke V., Massmann G., Hübner U. and Drewes J. E. (2018). Establishing sequential managed aquifer recharge technology (SMART) for enhanced removal of trace organic chemicals: experiences from field studies in Berlin, Germany. *Journal of Hydrology*, **563**, 1161–1168. https://doi.org/10.1016/j.jhydrol.2017.09.044

Henzler A. F., Greskowiak J. and Massmann G. (2014). Modeling the fate of organic micropollutants during river bank filtration (Berlin, Germany). *Journal of Contaminant Hydrology*, **156**, 78–92. https://doi.org/10.1016/j.jconhyd.2013.10.005

Hillebrand O., Nödler K., Licha T., Sauter M. and Geyer T. (2012). Caffeine as an indicator for the quantification of untreated wastewater in karst systems. *Water Research*, **46**(2), 395–402. https://doi.org/10.1016/j.watres.2011.11.003

Hrkal Z., Eckhardt P., Hrabánková A., Novotná E. and Rozman D. (2018). PPCP Monitoring in drinking water supply systems: the example of Káraný waterworks in Central Bohemia. *Water (Switzerland)*, **10**(12), 1–13. https://doi.org/10.3390/w10121852

Hu F., Dong F., Yin L., Wang H., Zheng M., Fu S. and Zhang W. (2021). Effects of sulfamethoxazole on the growth, oxidative stress and inflammatory response in the liver of juvenile Nile tilapia (Oreochromis niloticus). *Aquaculture*, **543**, 1–8. https://doi.org/10.1016/j.aquaculture.2021.736935

Hübner U., Seiwert B., Reemtsma T. and Jekel M. (2014). Ozonation products of carbamazepine and their removal from secondary effluents by soil aquifer treatment – indications from column experiments. *Water Research*, **49**(1), 34–43. https://doi.org/10.1016/j.watres.2013.11.016

Jelic A., Fatone F., di Fabio S., Petrovic M., Cecchi F. and Barcelo D. (2012). Tracing pharmaceuticals in a municipal plant for integrated wastewater and organic solid waste treatment. *Science of the Total Environment*, **433**, 352–361. https://doi.org/10.1016/j.scitotenv.2012.06.059

Jiang Y., Li M., Guo C., An D., Xu J., Zhang Y. and Xi B. (2014). Distribution and ecological risk of antibiotics in a typical effluent-receiving river (Wangyang River) in north China. *Chemosphere*, **112**, 267–274. https://doi.org/10.1016/j.chemosphere.2014.04.075

Joint Research Centre (2003). Commission E. Technical Guidance Document (TGD) in support of Commission Directive 93/67/EEC on Risk Assessment for new notified substances, Commission Regulation (EC) No 1488/94 on Risk Assessment for existing substances and Directive 98/8/EC of the 44 European Parliament and the Council concerning the placing of biocidal products on the market. Patent No. PART II.

Jurado A., López-Serna R., Vázquez-Suné E., Carrera J., Pujades E., Petrovic M. and Barceló D. (2014). Occurrence of carbamazepine and five metabolites in an urban aquifer. *Chemosphere*, **115**(1), 47–53. https://doi.org/10.1016/j.chemosphere.2014.01.014

Jurado A., Margareto A., Pujades E., Vázquez-Suñé E. and Diaz-Cruz M. S. (2020). Fate and risk assessment of sulfonamides and metabolites in urban groundwater. *Environmental Pollution*, **267**, 1–11. https://doi.org/10.1016/j.envpol.2020.115480

Kar S. and Roy K. (2010). First report on interspecies quantitative correlation of ecotoxicity of pharmaceuticals. *Chemosphere*, **81**(6), 738–747. https://doi.org/10.1016/j.chemosphere.2010.07.019

Khan M. T., Shah I. A., Ihsanullah I., Naushad M., Ali S., Shah S. H. A. and Mohammad A. W. (2021). Hospital wastewater as a source of environmental contamination: an overview of management practices, environmental risks, and treatment processes. *Journal of Water Process Engineering*, **41**, 1–17. https://doi.org/10.1016/j.jwpe.2021.101990

Kibuye F. A., Gall H. E., Elkin K. R., Ayers B., Veith T. L., Miller M., Jacob S., Hayden K. R., Watson J. E. and Elliott H. A. (2019). Fate of pharmaceuticals in a spray-irrigation system: from wastewater to groundwater. *Science of the Total Environment*, **654**, 197–208. https://doi.org/10.1016/j.scitotenv.2018.10.442

Kiecak A., Sassine L., Boy-Roura M., Elsner M., Mas-Pla J., le Gal La Salle C. and Stumpp C. (2019). Sorption properties and behaviour at laboratory scale of selected pharmaceuticals using batch experiments. *Journal of Contaminant Hydrology*, **225**, 1–11. https://doi.org/10.1016/j.jconhyd.2019.103500

Kiecak A., Breuer F. and Stumpp C. (2020). Column experiments on sorption coefficients and biodegradation rates of selected pharmaceuticals in three aquifer sediments. *Water (Switzerland)*, **12**(1), 1–19. https://doi.org/10.3390/w12010014

Kim Y., Choi K., Jung J., Park S., Kim P. G. and Park J. (2007). Aquatic toxicity of Acetaminophen, carbamazepine, cimetidine, diltiazem and six major sulfonamides, and their potential ecological risks in Korea. *Environment International*, **33**(3), 370–375. https://doi.org/10.1016/j.envint.2006.11.017

Koba O., Golovko O., Kodešová R., Klement A. and Grabic R. (2016). Transformation of atenolol, metoprolol, and carbamazepine in soils: The identification, quantification, and stability of the transformation products and further implications for the environment. *Environmental Pollution*, **218**, 574–585. https://doi.org/10.1016/j.envpol.2016.07.041

Kodešová R., Kočárek M., Klement A., Golovko O., Koba O., Fér M., Nikodem A., Vondráčková L., Jakšík O. and Grabic R. (2016). An analysis of the dissipation of pharmaceuticals under thirteen different soil conditions. *Science of the Total Environment*, **544**, 369–381. https://doi.org/10.1016/j.scitotenv.2015.11.085

Kodešová R., Chroňáková A., Grabicová K., Kočárek M., Schmidtová Z., Frková Z., Vojs Staňová A., Nikodem A., Klement A., Fér M. and Grabic R. (2020). How microbial community composition, sorption and simultaneous application of six pharmaceuticals affect their dissipation in soils. *Science of the Total Environment*, **746**, 1–14. https://doi.org/10.1016/j.scitotenv.2020.141134

König A., Weidauer C., Seiwert B., Reemtsma T., Unger T. and Jekel M. (2016). Reductive transformation of carbamazepine by abiotic and biotic processes. *Water Research*, **101**, 272–280. https://doi.org/10.1016/j.watres.2016.05.084

K'oreje K. O., Okoth M., van Langenhove H. and Demeestere K. (2020). Occurrence and treatment of contaminants of emerging concern in the African aquatic environment: literature review and a look ahead. *Journal of Environmental Management*, **254**, 1–18. https://doi.org/10.1016/j.jenvman.2019.109752

Koroša A., Auersperger P. and Mali N. (2016). Determination of micro-organic contaminants in groundwater (Maribor, Slovenia). *Science of the Total Environment*, **571**, 1419–1431. https://doi.org/10.1016/j.scitotenv.2016.06.103

Koroša A., Brenčič M. and Mali N. (2020). Estimating the transport parameters of propyphenazone, caffeine and carbamazepine by means of a tracer experiment in a coarse-gravel unsaturated zone. *Water Research*, **175**, 1–12. https://doi.org/10.1016/j.watres.2020.115680

Kosjek T., Andersen H. R., Kompare B., Ledin A. and Heath E. (2009). Fate of carbamazepine during water treatment. *Environmental Science and Technology*, **43**(16), 6256–6261. https://doi.org/10.1021/es900070h

Kovalakova P., Cizmas L., McDonald T. J., Marsalek B., Feng M. and Sharma V. K. (2020). Occurrence and toxicity of antibiotics in the aquatic environment: a review. In: Chemosphere, Elsevier Ltd., **251**, 1–15. https://doi.org/10.1016/j.chemosphere.2020.126351

Kozel R., Wolter R., Amelin E. V., Grath J., Besien T., Bonte M., Claessens J., Delloye F., Eppinger R., Ferretti E., Hilliges F., Laabs V., Lapworth D., Lemke D., Lopez B. and Ward R. (2019). Voluntary Groundwater Watch List (VGWWL).

Kozub P., Kozub S., Mozaffari N., Sihag P., Mozaffari N., Vambol S., Vambol V., Ramsawak N. and Khan N. A. (2021). Treatment schemes – conventional and dedicated for PhACs treatment. In: Pharmaceutical Wastewater Treatment Technologies: Concepts and Implementation Strategies, IWA Publishing, London, pp. 181–220, https://doi.org/10.2166/9781789061338_0181

Lamichhane K., Garcia S. N., Huggett D. B., Deangelis D. L. and la Point T. W. (2013). Chronic effects of carbamazepine on life-history strategies of ceriodaphnia dubia in three successive generations. *Archives of Environmental Contamination and Toxicology*, **64**(3), 427–438. https://doi.org/10.1007/s00244-012-9845-5

Lapworth D. J., Baran N., Stuart M. E. and Ward R. S. (2012). Emerging organic contaminants in groundwater: a review of sources, fate and occurrence. *Environmental Pollution*, **163**, 1–17. https://doi.org/10.1016/j.envpol.2011.12.034

Lapworth D. J., Baran N., Stuart M. E., Manamsa K. and Talbot J. (2015). Persistent and emerging micro-organic contaminants in chalk groundwater of England and France. *Environmental Pollution*, **203**, 214–225. https://doi.org/10.1016/j.envpol.2015.02.030

Lapworth D. J., Das P., Shaw A., Mukherjee A., Civil W., Petersen J. O., Gooddy D. C., Wakefield O., Finlayson A., Krishan G., Sengupta P. and MacDonald A. M. (2018). Deep urban groundwater vulnerability in India revealed through the use of emerging organic contaminants and residence time tracers. *Environmental Pollution*, **240**, 938–949. https://doi.org/10.1016/j.envpol.2018.04.053

Lapworth D. J., Lopez B., Laabs V., Kozel R., Wolter R., Ward R., Vargas Amelin E., Besien T., Claessens J., Delloye F., Ferretti E. and Grath J. (2019). Developing a groundwater watch list for substances of emerging concern: A European perspective. *Environmental Research Letters*, **14**(3), 1–15. https://doi.org/10.1088/1748-9326/aaf4d7

le Corre K., Katouli M., Stratton H., Ort C. and Keller J. (2012). Hospital Wastewater, http://www.griffith.edu.au/ (accessed 30 August 2021)

Lesser L. E., Mora A., Moreau C., Mahlknecht J., Hernández-Antonio A., Ramírez A. I. and Barrios-Piña H. (2018). Survey of 218 organic contaminants in groundwater derived from the world's largest untreated wastewater irrigation system: Mezquital Valley, Mexico. *Chemosphere*, **198**, 510–521. https://doi.org/10.1016/j.chemosphere.2018.01.154

Li Z. H., Zlabek V., Velisek J., Grabic R., Machova J., Kolarova J., Li P. and Randak T. (2011). Acute toxicity of carbamazepine to juvenile rainbow trout (Oncorhynchus mykiss): effects on antioxidant responses, hematological parameters and hepatic EROD. *Ecotoxicology and Environmental Safety*, **74**(3), 319–327. https://doi.org/10.1016/j.ecoenv.2010.09.008

Li J., Dodgen L., Ye Q. and Gan J. (2013). Degradation kinetics and metabolites of carbamazepine in soil. *Environmental Science and Technology*, **47**(8), 3678–3684. https://doi.org/10.1021/es304944c

Lin K. and Gan J. (2011). Sorption and degradation of wastewater-associated non-steroidal anti-inflammatory drugs and antibiotics in soils. *Chemosphere*, **83**(3), 240–246. https://doi.org/10.1016/j.chemosphere.2010.12.083

Lin Y. C., Lai W. W. P., Tung H. h. and Lin A. Y. C. (2015). Occurrence of pharmaceuticals, hormones, and perfluorinated compounds in groundwater in Taiwan. *Environmental Monitoring and Assessment*, **187**(5), 1–19. https://doi.org/10.1007/s10661-015-4497-3

Liu F., Ying G. G., Yang J. F., Zhou L. J., Tao R., Wang L., Zhang L. J. and Peng P. A. (2010). Dissipation of sulfamethoxazole, trimethoprim and tylosin in a soil under aerobic and anoxic conditions. *Environmental Chemistry*, **7**(4), 370–376. https://doi.org/10.1071/EN09160

Liu J., Wei T., Wu X., Zhong H., Qiu W. and Zheng Y. (2020). Early exposure to environmental levels of sulfamethoxazole triggers immune and inflammatory response of healthy zebrafish larvae. *Science of the Total Environment*, **703**, 1–10. https://doi.org/10.1016/j.scitotenv.2019.134724

Loos R., Locoro G., Comero S., Contini S., Schwesig D., Werres F., Balsaa P., Gans O., Weiss S., Blaha L., Bolchi M. and Gawlik B. M. (2010). Pan-European survey on the occurrence of selected polar organic persistent pollutants in ground water. *Water Research*, **44**(14), 4115–4126. https://doi.org/10.1016/j.watres.2010.05.032

Lopez B., Ollivier P., Togola A., Baran N. and Ghestem J. P. (2015). Screening of French groundwater for regulated and emerging contaminants. *Science of the Total Environment*, **518–519**, 562–573. https://doi.org/10.1016/j.scitotenv.2015.01.110

López-Serna R., Jurado A., Vázquez-Suñé E., Carrera J., Petrović M. and Barceló D. (2013). Occurrence of 95 pharmaceuticals and transformation products in urban groundwaters underlying the metropolis of Barcelona, Spain. *Environmental Pollution*, **174**, 305–315. https://doi.org/10.1016/j.envpol.2012.11.022

Ma Y., Li M., Wu M., Li Z. and Liu X. (2015). Occurrences and regional distributions of 20 antibiotics in water bodies during groundwater recharge. *Science of the Total Environment*, **518–519**, 498–506. https://doi.org/10.1016/j.scitotenv.2015.02.100

Ma L., Liu Y., Zhang J., Yang Q., Li G. and Zhang D. (2018). Impacts of irrigation water sources and geochemical conditions on vertical distribution of pharmaceutical and personal care products (PPCPs) in the vadose zone soils. *Science of the Total Environment*, **626**, 1148–1156. https://doi.org/10.1016/j.scitotenv.2018.01.168

McCance W., Jones O. A. H., Edwards M., Surapaneni A., Chadalavada S. and Currell M. (2018). Contaminants of emerging concern as novel groundwater tracers for delineating wastewater impacts in urban and peri-urban areas. *Water Research*, **146**, 118–133. https://doi.org/10.1016/j.watres.2018.09.013

McEachran A. D., Shea D., Bodnar W. and Nichols E. G. (2016). Pharmaceutical occurrence in groundwater and surface waters in forests land-applied with municipal wastewater. *Environmental Toxicology and Chemistry*, **35**(4), 898–905. https://doi.org/10.1002/and so on.3216

Müller B., Scheytt T., Asbrand M. and de Casas A. M. (2012). Pharmaceuticals as indictors of sewage-influenced groundwater. *Hydrogeology Journal*, **20**(6), 1117–1129. https://doi.org/10.1007/s10040-012-0852-4

Mutiyar P. K. and Mittal A. K. (2014). Risk assessment of antibiotic residues in different water matrices in India: key issues and challenges. *Environmental Science and Pollution Research*, **21**(12), 7723–7736. https://doi.org/10.1007/s11356-014-2702-5

Nakada N., Shinohara H., Murata A., Kiri K., Managaki S., Sato N. and Takada H. (2007). Removal of selected pharmaceuticals and personal care products (PPCPs) and endocrine-disrupting chemicals (EDCs) during sand filtration and ozonation at a municipal sewage treatment plant. *Water Research*, **41**(19), 4373–4382. https://doi.org/10.1016/j.watres.2007.06.038

Oberleitner D., Schulz W., Bergmann A. and Achten C. (2020). Impact of seasonality, redox conditions, travel distances and initial concentrations on micropollutant removal during riverbank filtration at four sites. *Chemosphere*, **250**, 1–10. https://doi.org/10.1016/j.chemosphere.2020.126255

Park S. and Lee W. (2018). Removal of selected pharmaceuticals and personal care products in reclaimed water during simulated managed aquifer recharge. *Science of the Total Environment*, **640–641**, 671–677. https://doi.org/10.1016/j.scitotenv.2018.05.221

Peake B. M., Braund R., Tong A. Y. C. and Tremblay L. A. (2016). Impact of pharmaceuticals on the environment. In: The Life-Cycle of Pharmaceuticals in the Environment, Peake B. M., Braund R., Tong A. and Tremblay L. (eds), Elsevier, Woodhead Publishing, London, pp. 109–152, https://doi.org/10.1016/b978-1-907568-25-1.00005-0

Pereira L. C., de Souza A. O., Bernardes M. F. F., Pazin M., Tasso M. J., Pereira P. H. and Dorta D. J. (2015). A perspective on the potential risks of emerging contaminants to human and environmental health. *Environmental Science and Pollution Research*, **22**(18), 13800–13823. https://doi.org/10.1007/s11356-015-4896-6

Petrović M., Škrbić B., Živančev J., Ferrando-Climent L. and Barcelo D. (2014). Determination of 81 pharmaceutical drugs by high performance liquid chromatography coupled to mass spectrometry with hybrid triple quadrupole-linear ion trap in different types of water in Serbia. *Science of the Total Environment*, **468–469**, 415–428. https://doi.org/10.1016/j.scitotenv.2013.08.079

Piersma A., Verhoef A., Opperhuizen A., Klaassen R., van Eijkeren J. and Olling M. (1998). Embriotoxicity of carbamazepine in rat postimplantation embryo culture after in vitro exposure via three different routes. *Reproductive Toxicology*, **12**(2), 161–168, https://doi.org/10.1016/S0890-6238(97)00152-4

Qin L. T., Pang X. R., Zeng H. H., Liang Y. P., Mo L. Y., Wang D. Q. and Dai J. F. (2020). Ecological and human health risk of sulfonamides in surface water and groundwater of Huixian karst wetland in Guilin, China. *Science of the Total Environment*, **708**, 1–9. https://doi.org/10.1016/j.scitotenv.2019.134552

Reh R., Licha T., Geyer T., Nödler K. and Sauter M. (2013). Occurrence and spatial distribution of organic micro-pollutants in a complex hydrogeological karst system during low flow and high flow periods, results of a two-year study. *Science of the Total Environment*, **443**, 438–445. https://doi.org/10.1016/j.scitotenv.2012.11.005

Rodríguez-Escales P. and Sanchez-Vila X. (2016). Fate of sulfamethoxazole in groundwater: conceptualizing and modeling metabolite formation under different redox conditions. *Water Research*, **105**, 540–550. https://doi.org/10.1016/j.watres.2016.09.034

Rozman D., Hrkal Z., Váňa M., Vymazal J. and Boukalová Z. (2017). Occurrence of pharmaceuticals in wastewater and their interaction with shallow aquifers: a case study of Horní Beřkovice, Czech Republic. *Water (Switzerland)*, **9**(3), 1–15. https://doi.org/10.3390/w9030218

Sacher F., Thomas Lange F., Brauch H.-J. and Blankenhorn I. (2001). Pharmaceuticals in groundwaters analytical methods and results of a monitoring program in Baden-Wurttemberg, Germany. *Journal of Chromatography A*, **938**, 199–210, www.elsevier.com/locate/chroma; https://doi.org/10.1016/S0021-9673(01)01266-3

Schaffer M., Boxberger N., Börnick H., Licha T. and Worch E. (2012). Sorption influenced transport of ionizable pharmaceuticals onto a natural sandy aquifer sediment at different pH. *Chemosphere*, **87**(5), 513–520. https://doi.org/10.1016/j.chemosphere.2011.12.053

Silver M., Selke S., Balsaa P., Wefer-Roehl A., Kübeck C. and Schüth C. (2018). Fate of five pharmaceuticals under different infiltration conditions for managed aquifer recharge. *Science of the Total Environment*, **642**, 914–924. https://doi.org/10.1016/j.scitotenv.2018.06.120

Sousa J. C. G., Ribeiro A. R., Barbosa M. O., Pereira M. F. R. and Silva A. M. T. (2018). A review on environmental monitoring of water organic pollutants identified by EU guidelines. *Journal of Hazardous Materials*. Elsevier B.V., **344**, 146–162, https://doi.org/10.1016/j.jhazmat.2017.09.058

Sposito G. (2008). The Chemistry of Soils, 2nd edn. Oxford University Press, Oxford.

Srinivasan P. and Sarmah A. K. (2014). Assessing the sorption and leaching behaviour of three sulfonamides in pasture soils through batch and column studies. *Science of the Total Environment*, **493**, 535–543. https://doi.org/10.1016/j.scitotenv.2014.06.034

Straub J. O. (2016). Aquatic environmental risk assessment for human use of the old antibiotic sulfamethoxazole in Europe. *Environmental Toxicology and Chemistry*, **35**(4), 767–779. https://doi.org/10.1002/and so on.2945

Strauss C., Harter T. and Radke M. (2011). Effects of pH and manure on transport of sulfonamide antibiotics in soil. *Journal of Environmental Quality*, **40**(5), 1652–1660. https://doi.org/10.2134/jeq2010.0535

Stuart M., Lapworth D., Crane E. and Hart A. (2012). Review of risk from potential emerging contaminants in UK groundwater. *Science of the Total Environment*, **416**, 1–21. https://doi.org/10.1016/j.scitotenv.2011.11.072

Stuart M. E., Lapworth D. J., Thomas J. and Edwards L. (2014). Fingerprinting groundwater pollution in catchments with contrasting contaminant sources using microorganic compounds. *Science of the Total Environment*, **468–469**, 564–577. https://doi.org/10.1016/j.scitotenv.2013.08.042

Teijon G., Candela L., Tamoh K., Molina-Díaz A. and Fernández-Alba A. R. (2010). Occurrence of emerging contaminants, priority substances (2008/105/CE) and heavy metals in treated wastewater and groundwater at Depurbaix facility (Barcelona. *Spain). Science of the Total Environment*, **408**(17), 3584–3595. https://doi.org/10.1016/j.scitotenv.2010.04.041

Vambol S., Vambol V., Mozaffari N., Mahmoud A. E. D., Ramsawak N., Mozaffari N., Ziarati P. and Khan N. A. (2021). Comprehensive insights into sources of pharmaceutical wastewater in the biotic systems. In: Pharmaceutical Wastewater Treatment Technologies: Concepts and Implementation Strategies, Khan N. A., Ahmed S., Vambol V. and Vambol S. (eds), IWA Publishing, London, pp. 17–46, https://doi.org/10.2166/9781789061338_0017

van den Brandhof E. J. and Montforts M. (2010). Fish embryo toxicity of carbamazepine, diclofenac and metoprolol. *Ecotoxicology and Environmental Safety*, **73**(8), 1862–1866. https://doi.org/10.1016/j.ecoenv.2010.08.031

van Driezum I. H., Derx J., Oudega T. J., Zessner M., Naus F. L., Saracevic E., Kirschner A. K. T., Sommer R., Farnleitner A. H. and Blaschke A. P. (2019). Spatiotemporal resolved sampling for the interpretation of micropollutant removal during riverbank filtration. *Science of the Total Environment*, **649**, 212–223. https://doi.org/10.1016/j.scitotenv.2018.08.300

van Stempvoort D. R., Roy J. W., Grabuski J., Brown S. J., Bickerton G. and Sverko E. (2013). An artificial sweetener and pharmaceutical compounds as co-tracers of urban wastewater in groundwater. *Science of the Total Environment*, **461–462**, 348–3359. https://doi.org/10.1016/j.scitotenv.2013.05.001

Ven A. J. A. M. van der, Vree T. B., J Van Ewijk-beneken Kolmer E. W., Koopmans P. P. and M Van Der Meer J. W. (1995). Urinary recovery and kinetics of sulphamethoxazole and its metabolites in HIV-seropositive patients and healthy volunteers after a single oral dose of sulphamethoxazole. *British Journal of Clinical Pharmacology*, **39**, 621–625, https://doi.org/10.1111/j.1365-2125.1995.tb05721.x

von der Ohe P. C., Dulio V., Slobodnik J., de Deckere E., Kühne R., Ebert R. U., Ginebreda A., de Cooman W., Schüürmann G. and Brack W. (2011). A new risk assessment approach for the prioritization of 500 classical and emerging organic microcontaminants as potential river basin specific pollutants under the European Water Framework Directive. *Science of the Total Environment*, **409**(11), 2064–2077. https://doi.org/10.1016/j.scitotenv.2011.01.054

Vulliet E. and Cren-Olivé C. (2011). Screening of pharmaceuticals and hormones at the regional scale, in surface and groundwaters intended to human consumption. *Environmental Pollution*, **159**(10), 2929–2934. https://doi.org/10.1016/j.envpol.2011.04.033

Wang D., Zhang X. and Yan C. (2019). Occurrence and removal of sulfonamides and their acetyl metabolites in a biological aerated filter (BAF) of wastewater treatment plant in Xiamen, south China. *Environmental Science and Pollution Research*, **26**(32), 33363–33372. https://doi.org/10.1007/s11356-019-06311-7

Wiese B., Massmann G., Jekel M., Heberer T., Dünnbier U., Orlikowski D. and Grützmacher G. (2011). Removal kinetics of organic compounds and sum parameters under field conditions for managed aquifer recharge. *Water Research*, **45**(16), 4939–4950. https://doi.org/10.1016/j.watres.2011.06.040

Williams C. F. and McLain J. E. T. (2012). Soil persistence and fate of carbamazepine, lincomycin, caffeine, and ibuprofen from wastewater reuse. *Journal of Environmental Quality*, **41**(5), 1473–1480. https://doi.org/10.2134/jeq2011.0353

Wolf L., Zwiener C. and Zemann M. (2012). Tracking artificial sweeteners and pharmaceuticals introduced into urban groundwater by leaking sewer networks. *Science of the Total Environment*, **430**, 8–19. https://doi.org/10.1016/j.scitotenv.2012.04.059

Xiang J., Wu M., Lei J., Fu C., Gu J. and Xu G. (2018). The fate and risk assessment of psychiatric pharmaceuticals from psychiatric hospital effluent. *Ecotoxicology and Environmental Safety*, **150**, 289–296. https://doi.org/10.1016/j.ecoenv.2017.12.049

Yan Z., Yang Q., Jiang W., Lu J., Xiang Z., Guo R. and Chen J. (2018). Integrated toxic evaluation of sulfamethazine on zebrafish: including two lifespan stages (embryo-larval and adult) and three exposure periods (exposure, post-exposure and re-exposure). *Chemosphere*, **195**, 784–792. https://doi.org/10.1016/j.chemosphere.2017.12.119

Yang Y. Y., Zhao J. L., Liu Y. S., Liu W. R., Zhang Q. Q., Yao L., Hu L. X., Zhang J. N., Jiang Y. X. and Ying G. G. (2018). Pharmaceuticals and personal care products (PPCPs) and artificial sweeteners (ASs) in surface and ground waters and their application as indication of wastewater contamination. *Science of the Total Environment*, **616–617**, 816–823. https://doi.org/10.1016/j.scitotenv.2017.10.241

Yu J. T., Bouwer E. J. and Coelhan M. (2006). Occurrence and biodegradability studies of selected pharmaceuticals and personal care products in sewage effluent. *Agricultural Water Management*, **86**(1–2), 72–80. https://doi.org/10.1016/j.agwat.2006.06.015

Yu Y., Liu Y. and Wu L. (2013). Sorption and degradation of pharmaceuticals and personal care products (PPCPs) in soils. *Environmental Science and Pollution Research*, **20**(6), 4261–4267. https://doi.org/10.1007/s11356-012-1442-7

Zainab S. M., Junaid M., Xu N. and Malik R. N. (2020). Antibiotics and antibiotic resistant genes (ARGs) in groundwater: a global review on dissemination, sources, interactions, environmental and human health risks. *Water Research*, **187**, 1–17. https://doi.org/10.1016/j.watres.2020.116455

Zainab S. M., Junaid M., Rehman M. Y. A., Lv M., Yue L., Xu N. and Malik R. N. (2021). First insight into the occurrence, spatial distribution, sources, and risks assessment of antibiotics in groundwater from major urban-rural settings of Pakistan. *Science of the Total Environment*, **791**, 1–11. https://doi.org/10.1016/j.scitotenv.2021.148298

Zemann M., Wolf L., Pöschko A., Schmidt N., Sawarieh A., Seder N., Tiehm A., Hötzl H. and Goldscheider N. (2014). Sources and processes affecting the spatio-temporal distribution of pharmaceuticals and X-ray contrast media in the water resources of the Lower Jordan Valley, Jordan. *Science of the Total Environment*, **488–489**(1), 100–114. https://doi.org/10.1016/j.scitotenv.2014.04.063

Zentner E., Gerstl Z., Weisbrod N., Lev O., Pankratov I., Russo D., Gasser G., Voloshenko Rosin A. and Ronen D. (2015). Deep penetration of pharmaceuticals and personal care products through the Vadose zone of Effluent-Irrigated land. *Vadose Zone Journal*, **14**(1), 1–7. https://doi.org/10.2136/vzj2014.09.0115

Zhang Y., Geißen S. U. and Gal C. (2008). Carbamazepine and diclofenac: removal in wastewater treatment plants and occurrence in water bodies. *Chemosphere*, **73**(8), 1151–1161. https://doi.org/10.1016/j.chemosphere.2008.07.086

Zorginstituut Nederland (2018). *Top 500 of medicines based on the number of DDDs in 2020*. GIP Database, https://www.gipdatabank.nl/databank?infotype=g&label=00-totaal&tabel=R_46_top500_atclaatst&geg=ddd&item= (accessed 30 October 2021)

Zuo R., Liu X., Zhang Q., Wang J., Yang J., Teng Y., Chen X. and Zhai Y. (2021). Sulfonamide antibiotics in groundwater and their migration in the vadose zone: a case in a drinking water resource. *Ecological Engineering*, **162**, 1–2. https://doi.org/10.1016/j.ecoleng.2021.106175

doi: 10.2166/9781789062625_0133

Chapter 7

Case studies of hospital effluent in Ukraine urban treatment plants

Kravets Natalia and Trach Iryna

Vinnytsia National Technical University, Vinnytsia, Ukraine
Corresponding authors. trachiryna2103@gmail.com, kravets19950401@gmail.com

ABSTRACT

The ecological analysis of sources of pharmaceutical pollution of the aquatic environment is carried out in this work. A detailed classification of water pollution by pharmaceuticals, including clogging of reservoirs is presented. An analysis of pollution of the water environment by branches of the medical industry, and their influence on the condition of reservoirs is carried out. The principles of assessment of pollutants in natural waters are considered. Drugs are created to affect the human body, but due to the similarity of physiological mechanisms in different species, drugs can also affect other organisms, in ecosystems at the individual, species and interspecific levels. In this regard, the study of the life cycle of drugs in the environment and elucidation of the effects of drug residues on various living organisms is a field of research in environmental toxicology.

Keywords: ecological analysis, water environment, pharmaceutical industry, pollution

CONTENTS

7.1 INTRODUCTION

In the last two decades, pharmaceuticals and their metabolites have increasingly been detected in the environment, becoming a chemically diverse class of new organic pollutants. All living organisms in the environment are, to some extent, under the influence of drugs present in it, so even relatively low concentrations of drugs can have a significant impact on the state of the ecosystem.

Editor(s): Nadeem A Khan, Viola Vambol, Sergij Vambol, Nastaran Mozaffari and Niloofar Mozaffari

Drugs in drinking water are a significant global environmental threat. Pharmaceutical companies do not want to hear about this problem – they do not want to spend huge amounts of money on the development and implementation of drugs that would be fully absorbed by the body. But in the next two or three years, they will have to listen to environmentalists and radically change the strategy of drug production.

Antibiotic contamination is one of the key ways in which antibiotic resistance is developed, leading to the loss of their effectiveness in treatment. Resistance to antibiotics, antivirals, antifungals and antiprotozoals is a global problem. If left untreated, 10 million people could die from antibiotic resistance by 2050.

The pharmaceutical industry, which increases the production of medicines by several million packages annually and continually expands their range, significantly pollutes the environment. The environmental safety of such industries is regulated by law, so significant emissions are not systematic.

Due to the consistent increase of manufacturability and organization of the production process, the introduction of higher quality and environmental safety standards, and control by authorized state bodies in recent years, there has been an overall reduction in the environmental burden from pharmaceutical industries, especially in developed countries, in recent years. In addition, pharmaceutical production is localized geographically, and in the event of an accident or violation of environmental legislation, emissions are purely local and are hazardous only to a specific region.

Numerous studies have proven the extremely negative impact on water bodies of a small number of drugs that enter them with wastewater. Groups of drugs that are poorly soluble in water, including anti-inflammatory drugs, antibiotics, hormones, and drugs that lower cholesterol, have been studied. They are poorly biodegradable and, on passing through treatment plants unchanged, are released into natural waters.

The main reasons for the accumulation of pharmaceutical household waste are: the availability of medicines; advertising of pharmaceutical companies aimed at increasing drug consumption; polypragmatism; self-medication and self-diagnosis; old age and/or the presence of chronic diseases in people living in the house; and the presence of small children in the family.

Accumulated medicines become unusable due to damage to the original packaging, improper storage or exceeding the expiration date. After that, they are often either flushed down the drain or thrown away as household waste.

Thousands of pharmaceutical factories annually release tens of millions of tons of drugs to the world market. Sales of medicines in the world exceed sales of oil. Until recently, no one thought about their final fate. Scientists have found that more than half of drugs taken are excreted from the human body in a biologically active form with almost no loss of their properties. Huge quantities of these find their way into wastewater, and from there into drinking water sources.

A large number of medicines are not used by the population and end up in garbage containers. Drugs and their metabolites are increasingly found in the aquatic environment. Modern chemical pharmaceutical substances are very complex and pure substances, so many different chemical components are used in their production. And all of this, in one way or another, gets into drinking water.

7.2 ESTIMATION OF POLLUTANT LOAD ASSOCIATED WITH PHARMACEUTICALS RELEASED TO THE ENVIRONMENT

In recent decades, the problem of the growing uncontrolled presence of drug residues and their metabolites in the environment has attracted the attention of many researchers around the world (Vambol *et al.*, 2021a).

The question of the potential danger of such substances for biological species began to be studied in the 1990s. It is important to consider the presence, compliance, dissemination, detection methods, biodegradation, means to prevent blocking, and notification of removal from the environment, of the remains of various pharmaceuticals (Bespamyatnov & Krotov, 1985).

The pharmaceutical industry increases the production of medicines by several million packages annually with a continually expanding range and, historically, has placed a significant pollution burden on the environment. The environmental safety of such industries is regulated by law, so significant emissions are not systematic.

The study of advanced technology and organization of the production process, the introduction of higher standards of quality and environmental safety, and control by orderly government agencies in recent years, has contributed to the effort to reduce the environmental load from pharmaceutical industries, especially in developed countries. In addition, pharmaceutical production is localized geographically, and in the case of accidents or violations of environmental legislation, emissions are local in nature and are dangerous only for a particular region (Vishnevetsky & Vishnevetsky, 2004).

At this time, the problem of disposal of substandard drugs and drugs which have exceeded their expiration date is acute. In accordance with the order of the Ministry of Health of Ukraine No 349 of 08.07.2004, medicinal waste can be disposed of by using as secondary materials or energy resources in their own institution or transferred for disposal to another institution. The choice of method of disposal of waste medicines is determined in accordance with the requirements of state sanitary norms, taking into account their danger to public health and the environment. Disposal of drug waste is carried out after determination of the hazard class. Thus, wastes belonging to the 3rd hazard class can be disposed of in landfills for solid waste, and wastes of the 2nd safety class are subject to disposal by special methods.

Unfortunately, pharmacies and hospitals practice the repeated introduction and release of low-quality injectable forms of drugs of the 2nd and 3rd classes of hazard into the city sewerage system. Their decomposition in treatment plants is about 68%. More dangerous for the environment are the sources of medicines which are practically uncontrolled and come mainly from people who take medicines as part of their own medical care and also from animal husbandry.

A small proportion of drugs enter the aquatic environment due to their transport through the skin. Some drugs can be released under the influence of open water, which provides pollution of the aquatic environment.

Physiological secretions of animals are another necessary source of uncontrolled absorption of drugs in the environment. An example is the use of antibiotics in food for farm animals.

Various methods can be used to treat wastewater from pharmaceuticals. In most cases, they can be divided into two groups:

(1) traditional methods: biological treatment, chemical treatment methods based on the use of one oxidant, physicochemical treatment methods based on the use of one type of exposure;
(2) advanced methods – methods of purification based on the combined action of several substances or factors.

In Europe, the use of antibiotics (Tsalikov, 2009) for growth stimulation is prohibited. This reduces large-scale problems, but does not completely eliminate them. Initially, the bulk of drugs that subsequently pollute the environment, are discharged via domestic wastewater pipes as a result of physiological excretion and/or flushing of unnecessary drugs into the sewer.

Treated effluents are discharged into surface waters and bring with them the remains of drugs that have not been removed by the treatment system or the natural degradation of pharmaceutical substances. Surface waters carry drugs to seashores and replenish aquifers. Soil fertilized with sewage sludge or manure from farm animals that have been treated with drugs or that interact with surface fresh water containing drug residues or pharmaceutical waste, is the medium through which pharmaceutical substances infiltrate groundwater. The ability of substances to infiltrate is determined by their physicochemical properties.

Groundwater and surface fresh water are sources of drinking water. Despite the high-quality purification of natural waters carried out in the developed countries of the world, the remnants of medicines in certain quantities enter the water supply network and are then re-ingested by people using the drinking water. This was confirmed by the Associated Press in March 2008, regarding the presence of drug residues, including antibiotics, anticonvulsants and psychotropic substances, in the drinking water of 24 major urban areas of the United States, home to more than 41 million people (Khetan & Collins, 2007).

Thus, due to the increasing availability of medicines and the general development of health care systems, the consumption of medicines for medical purposes is increasing and, as a result, their content in the environment is increasing. This process is uncontrolled and potentially dangerous to human health and other biological organisms.

Sources of toxic pollution of water bodies are considered to be subjects and objects of the environment, which directly enable the entry of toxic substances into aquatic ecosystems.

The international chemical database includes 56 million items, and every year 50 000 newly synthesized compounds are added to it. MPCs (maximum permissible concentrations) are determined for only 1% of potential drinking water contaminants, and control laboratories detect an even lower percentage of toxicants.

Medical waste recycling is a task that requires great attention and carries great risks associated with the management of this medical waste. But the process of disposing of medical waste can lead to the release of large quantities of hazardous substances, and burial at a landfill or incineration must be carried out in accordance with environmental regulations and rules of the law of Ukraine. The process of burning medical waste releases into the atmosphere heavy environmental pollutants: carcinogens – these are furans and dioxins from chlorine halogen compounds. Essentially, the burial of hazardous materials that contain heavy metals, will likely result exposure to bioaccumulative hazardous substances (Korosteleva, 2013).

In 2007, German chemists Thomas Heberer and Hans-Jürgen Stan discovered significant amounts of a cholesterol-lowering drug in Europe, in the groundwater of Germany and Switzerland and scientists soon found other drugs in rivers, lakes and deep aquifers. So, each of us has the potential to consume a wide variety of active drugs every day from our domestic taps.

Wastewater treatment plants are usually not adapted to capture or decompose both the drugs themselves and many hygiene products. Effective methods to treat wastewater include ozonation, photocatalytic destruction, membrane filtration, reverse osmosis, thermal disposal, activated carbon, and others. There is even the potential to create special toilets for cleaning and disposing of urine (in hospitals, clinics). However, the cost of this additional water treatment would be 5–20 euros per year per person.

Waste drugs and pharmaceuticals are called organic micro-pollutants (OMPs). These include pharmaceuticals, personal care products, and pesticides and are listed as priority pollutants by the US Environmental Protection Agency and are a priority for the European Environment Agency. The appearance of OMPs in vital sources, such as surface and groundwater, and, accordingly, in drinking water, in concentrations from ng L^{-1} to several μg L^{-1} has an extremely negative impact on water quality.

In 2015, the European Commission developed a checklist of substances to be monitored across the European Union – Directive 2015/495/EU of 20 March 2015. The checklist includes substances previously mentioned in Directive 39/2013/EC, namely diclofenac, the synthetic hormone 17-alpha-ethinyl estradiol (EE2), the natural hormone 17-betaestradiol (E2) and estrone (E1). The checklist of 10 substances/groups of substances also includes three macrolide antibiotics (azithromycin, clarithromycin and erythromycin), a number of pesticides (methiocarb, oxadiazone, imidacloprid, thiacloprid, thiamethoxam, clothialipridin, −4-methoxycinnamate) and an antioxidant (2,6-di-tert-butyl-4-methylphenol), which is used as a food additive. Unfortunately, there is almost no information on substances that were not included in the checklist of Directive 2015/495/EU due to their low content in water samples during monitoring.

Organic micro-pollutants can be of both natural and anthropogenic origin. For example, sources in Central Europe are:

- industrial wastewater;
- drains of agriculture and animal husbandry;
- landfills for household waste;
- domestic and hospital wastewater from which they can enter various water sources, as shown in Figure 7.1 (Kozub *et al.*, 2021).

In the vast majority of cases, existing wastewater treatment technologies are not designed to remove organic micro-pollutants, both for physicochemical reasons and due to their low content in the treated water. The latter circumstance

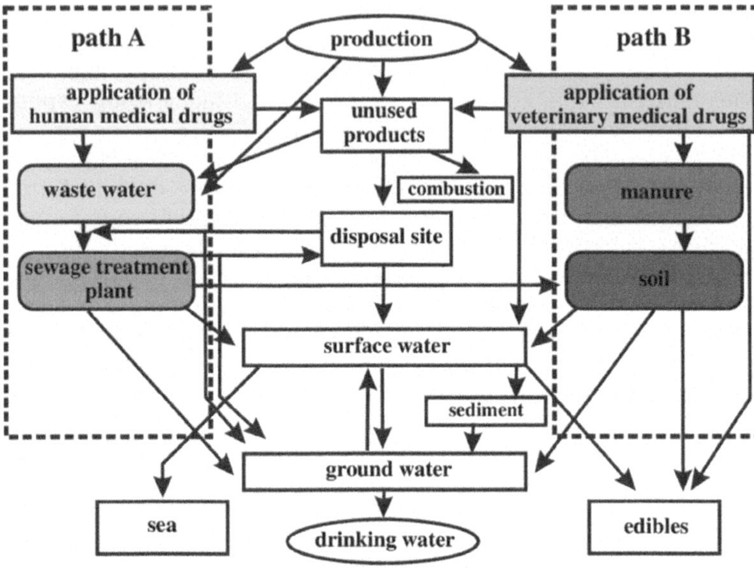

Figure 7.1 Sources of pharmaceuticals in the environment (Kozub *et al.*, 2021).

significantly reduces the effectiveness of traditional cleaning technologies, even in cases where there are no fundamental physical and chemical difficulties. At the same time, the toxicity of such compounds is very high and the presence of even minimal amounts of these substances in drinking water can lead to serious consequences (Maksyutina, 1976).

7.3 IMMEDIATE HEALTH RISK

The study of the mechanisms of entry of pharmaceuticals and their metabolites into the environment is a topical issue of modern science and attracts the attention of scientists from many countries (Khetan & Collins, 2007). In particular, the impact and assessment of the effects of pharmaceutical contaminants on wastewater, surface water and groundwater are studied.

To date, more than 3000 chemical substances have been found in Europe's water bodies.

The range of concentrations of pharmaceuticals in the environment is usually from ng L^{-1} to $\mu g\ L^{-1}$. Although this concentration is low and difficult to detect, it has a high potential for environmental effects.

The main source of pharmaceuticals and their metabolites in the environment are people who consume drugs. The first-aid kits of the developed countries of the world accumulate a large number of medicines that remain unused after completion of treatment or are purchased without direct need. Thus, a study conducted in the United States found that only 2% of outpatients complete treatment without residual drugs remaining. In another survey conducted in the UK, it was shown that 52.8% of all respondents complete treatment. Self-medication and non-compliance with treatment deadlines, as well as uncontrolled sale of pharmaceuticals in pharmacy chains, are prerequisites for the rapid accumulation of pharmaceutical waste, with no programs in many countries, including Ukraine, for its collection and disposal.

The main pollution of water with pharmaceuticals and their metabolites is due to wastewater from hospitals, clinics and pharmaceutical companies. Hospital wastewater has a 15 times higher ecotoxicity potential than urban wastewater.

The second largest emission of pharmaceuticals is agriculture, including livestock. Sources of entry into the environment of veterinary pharmaceuticals include prevention and treatment of diseases in animals using drugs, and organic fertilizers. A large number of antibacterial drugs are used in fisheries, in particular they are added to ponds for artificial breeding of fish.

The majority of waste pharmaceuticals together with sewage are transported to sewage treatment facilities where, depending on their properties, these substances, to some extent, are subject to destruction. Decomposition may not be complete, and the by-products formed may also be hazardous to the environment and humans. By-products of the destruction of pharmaceuticals can be both metabolites excreted from the body and biodegradation products of drugs – a variety of substances formed in the environment are from pharmaceuticals and/or their metabolites (Buryak & Lukash, 2012).

One of the factors influencing the quantitative and qualitative composition of pharmaceuticals entering the environment is their bioavailability. The higher it is, the lower the degree of excretion of the drug in its native state (Hurina, 2014).

Antibiotics are usually poorly metabolized in the body and excreted almost completely unchanged. Due to this, wastewater becomes a source of formation and spread of resistance factors. In accordance with the moderately low excretion of the starting compound, it is logical to assume the presence of metabolites in wastewater. Many pharmaceuticals form not one or two, but several dozen metabolites, the concentrations of which are relevant and important for environmental monitoring and enable not only the determination of the concentrations of native compounds, but also improvement of surface water treatment systems and other biosphere objects.

Contaminated water kills more people than wars and all other forms of violence combined. This was stated in a UN report published on the occasion of World Water Day, which has been celebrated annually on March 22 for the last 24 years.

Currently, more than 75% of the population of Ukraine consumes water from surface water sources, and providing the population of Ukraine with quality drinking water is becoming more difficult every year. According to the World Health Organization (WHO), about 80% of human diseases are related to the quality of drinking water. A glaring illustration of such an emergency was the poisoning of more than 150 people due to poor water quality in Kyiv's Bortnychi district in January last year. According to the Main Department of the Sanitary and Epidemiological Service in Kyiv, this happened due to the contamination of water with a pathogen of intestinal infection (Novikov *et al.*, 2008).

Experts note that about two billion (10^9) tons of wastewater, which is a mixture of recycled fertilizers, sewage, and industrial and human waste, is dumped into rivers and seas every day, spreading disease and causing irreparable damage to ecosystems. Therefore, the transformation of polluted water from a threat to health and the environment into a clean, safe and economically attractive resource is one of the key challenges for the world community in the 21st century.

The water we consume must be clean. Diseases transmitted through contaminated water cause deterioration in health, disability and death of a large number of people, especially children, mainly in less developed countries, where low levels of personal and communal hygiene are common. Diseases such as typhoid fever, dysentery, cholera and hookworm, are transmitted to humans primarily as a result of contamination of water sources with excrement excreted from the bodies of patients.

Success in the fight against these diseases, or achieving their complete elimination, depends on how organized the system of removal of all metabolic products excreted from the human body is, as part of the task of providing clean water to the entire population.

Infectious jaundice, tularemia, water fever, brucellosis, and polio can be transmitted through water. Water sometimes becomes a source of human infection with parasites. Eggs of some parasitic worms can get into the human body with water contaminated with feces. In the intestine, they turn into parasites (such as roundworms or pinworms). Finally, water sometimes causes infection by Giardia, which affects the small intestine and liver.

The largest polluters of surface water bodies continue to be industrial enterprises in the energy sector, ferrous metallurgy, chemistry and petrochemistry, utilities and agriculture.

Water quality is also determined by the presence of chemical inclusions in it, which are primarily detected by our senses: smell and sight. For example, microparticles of copper give the water some turbidity, and iron causes redness.

The presence of iron in water does not threaten our health. However, the high content of iron salts in water gives it an unpleasant swampy taste. If you wash clothes in such water, it will cause rusty stains. Similar stains appear on dishes, sinks and baths.

Occasionally there are many salts of hydrochloric and sulfuric acids (chlorides and sulfates) in drinking water. They give the water a salty and bitter-salty taste. The use of such water leads to disruption of the gastrointestinal tract. In water, chloride levels in excess of 350 mg L^{-1} and sulfate levels in excess of 500 mg L^{-1} are considered unfavorable for health.

Levels of salts of calcium and magnesium in water are closely related to another property of water, its hardness. Water highly saturated with salt causes a lot of inconvenience: it is harder to boil vegetables and meat; soap consumption for washing increases; scale spoils kettles and boilers; and clogs may form in water pipes. Studies have shown that there is a link between the use of hard water and the prevalence of some diseases (Strelnikov *et al.*, 2006).

This conclusion was reached by West German physicians who studied the composition of water and the prevalence of the most common diseases in various cities in Germany. It turned out that the more salts and impurities that were present in the water of a city, the lower the cases of heart attack and hypertension of the townspeople using this water.

The same opinion is held by English scientists. According to research by Dr. Thomas Grau Ford from London, Glasgow, where the water is very soft, has the highest mortality from cardiovascular disease in the British Isles. In London, the picture is completely different: the number of fatal heart attacks here is 37% less than in Glasgow.

But in addition to useful impurities in the water there are others which are dangerous to the human body. According to domestic researchers, the use of mine water containing 0.2–1 mg L^{-1} of arsenic causes a disorder of the central, and especially peripheral, nervous system with the subsequent development of polyneuritis. Arsenic is harmless in water at a concentration of 0.05 mg L^{-1}.

Hygienists first spoke about the health hazards of lead in water due to the mass intoxications that occurred when using lead pipes for water pipes. However, elevated concentrations of lead can occur in groundwater. Lead in water is considered harmless if its concentration does not exceed 0.03 mg L^{-1}.

Beryllium is quite common in nature. It is found in some natural waters. Beryllium is a poison of general toxic action, which can accumulate in the human body and then lead to damage to the respiratory, nervous and cardiovascular systems. The maximum allowable concentration of beryllium in drinking water is 0.002 mg L^{-1}.

Molybdenum is found in natural waters. Excessive exposure of the human body to molybdenum leads to molybdenum gout. A concentration of molybdenum in drinking water of 0.5 mg L^{-1} is considered harmless.

Strontium is widely distributed in natural waters, and its concentrations vary widely (from 0.1 to 45 mg L^{-1}). Prolonged intake in the body in large quantities leads to functional changes in the liver. However, long-term use of drinking water containing strontium at the level of 7 mg L^{-1} does not cause functional or morphological changes in tissues, organs and the whole human body. This value is accepted as the norm for strontium content in drinking water.

According to modern scientific data, nitrates in the human intestine, under the influence of bacteria living there, are reduced to nitrites. Absorption of nitrates leads to the formation of methemoglobin and to a partial loss of hemoglobin activity in oxygen transport.

Thus, methemoglobinemia is based on a degree of oxygen starvation, the symptoms of which are manifested primarily in children, especially infants. They get sick mainly during artificial feeding, when dry milk formulas are diluted with water containing nitrates, or when using this water for drinking. The use of water containing 2–11 mg L^{-1} of nitrates does not cause an increase in the level of methemoglobin in the blood, while the use of water with a concentration of 50–100 mg L^{-1} sharply increases this level. Methemoglobinemia is manifested by cyanosis, an increase in the content of methemoglobin in the blood, and a decrease in blood pressure. These symptoms have been registered not only in children but also in adults. A concentration level of nitrates in drinking water of 10 mg L^{-1} is harmless (Fedorenko *et al.*, 2006).

Uranium is a widespread radioactive element found in natural waters. Particularly high concentrations can be found in groundwater. The rationing of uranium is not based on its radioactive properties, but on its toxic effects as a chemical element. The permissible content of uranium in drinking water is 1.7 mg L^{-1}.

The maximum permissible concentration in water of certain additives used for water clarification (e.g., polyacrylamide and aluminum sulfate) is strictly regulated.

It is no exaggeration to say that high-quality water that meets sanitary and epidemiological requirements is one of the essential conditions for maintaining human health. For the period from 2010 to 2015 in Ukraine there was a tendency for an annual decrease in the volume of water abstraction from natural water bodies. In 2010, 13 215 million cubic meters of water were taken from natural water bodies, but in 2015 only 9699 million cubic meters were abstracted (i.e., a decrease of 3516 million cubic meters, or 26.4%). Compared to 2014, water abstraction decreased in 2015 by 1806 million cubic meters or 15.7%.

In recent years, the view of water has changed. Not only hygienists, but also biologists, engineers, builders, economists, and politicians have begun to talk about it more and more often. It is clear that the rapid development of social production and urban planning, the growth of material well-being, and the cultural level of the population constantly increase the need for water, forcing it to be used more carefully.

7.4 FULL-SCALE PLANT CASE STUDIES

Extensive use of antibiotics has led to the development of microorganisms which are resistant to common antibiotics. Until recently, it was considered that the main source of pharmaceutical substances (starting materials and metabolites) impacting the ecosystem were products of human life, but with the growth of industrialization of society and technological developments, an increasing role in environmental pollution is played by pharmaceutical companies.

The effluents of any pharmaceutical company are the combination of several sources containing various pollutants. This limits the use of standard methods and makes it necessary to search for comprehensive solutions that enable effective organization of the disposal process. Due to technical and economic difficulties, in most cases companies use the method of 'dilution' to achieve permissible concentrations, which allows them to meet the requirements of the law but in no way solves the problem of environmental pollution by drugs (Kika, 2020).

One of the specific methods of monitoring environmental pollution by drugs is bioindication. This determines the degree of pollution of geophysical environments by using living organisms known as bioindicators.

Bioindication is defined as operational monitoring of the environment based on observations of the condition and behavior of biological objects (plants, animals, etc.). Use of this method is becoming more widespread, as indicator plants have the following advantages:

- provide an opportunity to summarize biologically important data on the environment;
- are able to respond to short-term and volley emissions of toxicants;
- respond to the speed of changes in the environment;

- indicate the places of accumulation of pollutants and pathways of their migration;
- make it possible to develop estimates of the harmful effects of toxicants on humans and wildlife in the early stages, and to normalize the permissible load on ecosystems.

Secondary use of wastewater and increased use of wastewater for irrigation can lead to the accumulation of contaminants in the soil. But the potential phytotoxicological effect on higher plants has not been sufficiently assessed. The research described below aimed to examine the effect on the germination, development, growth and physiology of lettuce (Lactuca sativa) after exposure to nonsteroidal anti-inflammatory drugs (NSAIDs).

In a study by Rede *et al.* (2019) three commonly used pharmaceuticals from different therapeutic groups, paracetamol (PCT), ibuprofen (IBU) and amoxicillin (AMX) were selected for assessment of their possible phytotoxic effects on Lactuca seed germination, development and growth. L. sativa was chosen because it is a species identified by bioindication standards and recommended for ecotoxicological studies due to its ability to respond quickly to chemicals.

Lettuce seeds (Lactuca sativa) were stored in the dark at 4°C. Phytotoxicity studies were performed on certified artificial sandy soil, which was slightly acidic (pH = 5) without organic matter, with a water content of 50% and silica content of approximately 90%.

The process of studying the phytotoxic effect was as follows: 20 seeds of L. sativa were placed per 100 g of pre-dried sandy soil in Petri dishes.

Each seed was placed in a hole, covered with sand, and 10 ml of deionized water/contaminant was added to the top of the soil. The Petri dishes were sealed with parafilm and placed inside a germination chamber equipped with daylight fluorescent lamps for 120 h at 24 ± 2°C.

After five days of exposure, the length of the roots, shoots and leaves of the plants was measured and the percentage of damaged plants was calculated.

The ecotoxic effect of the three pharmaceutical preparations (IBU, AMX and PCT) was evaluated using L. sativa seeds, taking into account either their individual phytotoxicity or the toxic effect of different ratios of mixtures.

Studies of each individual pharmaceutical preparation were performed, covering a wide range of concentrations and their solubility in water. The solutions were prepared in deionized water and 100 g of artificial soil was contaminated with 10 ml of diluted solution.

The toxic effect of mixtures of PCT, IBU and AMX on germination and early growth of L. sativa seeds was evaluated. The sum of concentrations of pharmaceuticals was taken into account in all tested mixtures, although different proportions of pharmaceuticals were considered, in an attempt to establish their possible effect on plants.

Several binary combinations of the three pharmaceuticals were made. A triple mixture (MD) combining PCT, IBU and AMX at the same concentration was also tested. In addition, some pharmaceuticals were also tested at a concentration of 100 ng g^{-1}, which was considered the worst case scenario due to the solubility limit of IBU in aqueous solutions.

When L. sativa was treated with a mixture of AMX and IBU solutions, the four endpoints (% seed germination, root elongation, shoot length and leaf length) showed no significant statistical differences from the control. It turned out significant statistical differences existed between individual drugs and their combination.

Significant statistical differences were found between IBU and MAIII mixture. Root elongation showed statistical differences in AMX, namely for MAI and MAII mixtures (binary combinations). The effect was more pronounced for the mixture of MAI (median = 24.49 mm), which represented the largest number of IBU. In addition, an increase in the length of the shoots of the mixtures was observed compared to individual pharmaceuticals, which was significantly different from IBU.

The phytotoxic effect was more pronounced for the mixture of MAIII. All mixtures of AMX and IBU showed an increase in leaf length, which was statistically significantly different from IBU.

The effect of a mixture of AMX and PCT on L. sativa did not cause significant differences from the control for the four endpoints.

Studies of the mixture showed an increase in root elongation compared to individual pharmaceuticals (maximum increase of 41 and 85% compared to individual PCT and AMX, respectively).

In addition, the mixtures MBI and MBII (binary combinations) had significant statistical differences compared to the individual exposure to AMX, mixture MBIII also had significant statistical differences compared to the two individual drugs.

Shoot length had the highest increase in the mixture MBIII (median = 12.18 mm), which was statistically significant different from both individual pharmaceuticals. Furthermore, statistically significant differences in the shoot length were also observed in mixture MBI relative to AMX. In relation to leaf length, the mixtures showed an increase as well, being statistically significant different for MBII and MBIII relative to AMX.

Regarding the three mixtures of IBU and PCT no statistically significant differences were observed in any studied endpoint when compared to the control. In the germination percentage there were statistically significant differences between IBU and the control. Additionally, it can be said that the germination percentage increased in the mixtures (median = 95.0–97.5%) in relation to IBU (median = 75.0%). Root elongation tended to increase for all the mixtures (median = 19.96–25.56 mm) relative to PCT (median = 16.49 mm). The effect was more pronounced when high proportions of IBU were present in the mixture as was the case with MCIII, which showed statistically significant differences from PCT.

The same behavior was observed for leaf length and once again all the mixtures presented statistically significant differences relative to the individual exposure to IBU.

The combination of the three pharmaceuticals did not affect endpoints when results were compared to control, but there were statistically significant differences when compared with individual pharmaceuticals. Root elongation of L. sativa increased compared to the individual pharmaceuticals, showing statistically significant differences relative to AMX which corresponded to an increment of about 120%. The shoot length of the mixture presented an increase of 21%, 51% and 138% compared to PCT, AMX and IBU, respectively. Leaf length showed an identical tendency as shoot length, and statistically significant differences relative to IBU were noticed, corresponding to an increment of about 86%.

In general, for the same total concentration, an enhancement in the root elongation, and shoot and leaf length of the lettuce exposed to the pharmaceutical mixtures was observed relative to the individual pharmaceuticals. The inhibitory effect on early growth detected on exposure to isolated pharmaceuticals tends to be canceled in the mixtures, which might be due to interactions between compounds or even to the plant's own metabolism. It is known that the presence of isolated pharmaceuticals in the soil can induce oxidative stress in plants, and then their antioxidant defenses can be induced (Gómez-Oliván *et al.*, 2014), leading to different responses (phytotoxic effects) depending on the pharmaceutical, its concentration and the sensitivity of the plant species (Oliveira *et al.*, 2013). For instance, IBU acts by inhibiting cyclooxygenases, enzymes that are responsible for the catalysis of the prostaglandins synthesis. In higher plants, prostaglandins are involved in defense mechanisms, therefore the exposure to this pharmaceutical might compromise the response of plants to the oxidative stress induced by pollutants present in soils. Furthermore, PCT induces oxidative stress in cells, leading to an increase in the reactive oxygen species (ROS) produced, which might damage the antioxidant defense system of the plant (Paíga *et al.*, 2016). PCT, IBU and AMX are widely consumed pharmaceuticals and have been detected in soils. It is not expected that the environmental concentrations of these pharmaceuticals could induce toxicity in the germination and development of lettuce seeds. Under real environmental conditions it is more likely to find mixtures of contaminants that continuously change composition and concentrations over time (Vambol *et al.*, 2021b).

Thus, the accumulation of pollutants in plants primarily depends on the type of soil, physicochemical properties of pollutants and interactions, and the type of plants (Vambol *et al.*, 2021c).

7.5 CONCLUSIONS

The environmental safety of such industries is regulated by law, so significant emissions are not systematic. In addition, pharmaceutical production is localized geographically, and in the case of accidents or violations of environmental legislation, emissions are local in nature and are dangerous only for a particular region.

The study of the mechanisms of entry of pharmaceuticals and their metabolites into the environment is a topical issue of modern science and attracts the attention of scientists from many countries. In particular, the impact and assessment of the effects of pharmaceutical contaminants on wastewater, surface water and groundwater are studied. To date, more than 3000 chemical substances have been found in Europe's water bodies.

REFERENCES

Bespamyatnov G. P. and Krotov Y. A. (1985). Maximum Allowable Concentrations of Chemicals in the Environment. Directory. Chemistry, Lviv.

Buryak N. B. and Lukash S. V. (2012). Problemy zbyrannya, transportuvannya ta utylizatsiyi tverdykh pobutovykh vidkhodiv v ukrayini. *Naukovyy Visnyk NLTU Ukrayiny*, **22**(5), 82–90.

Fedorenko O. I., Bondar O. I. and Kudin A. V. (2006). Fundamentals of Ecology: Textbook. K.: Knowledge.

Gómez-Oliván L. M., Neri-Cruz N. and Galar-Martínez M. (2014). Binary mixtures of diclofenac with paracetamol, ibuprofen, naproxen, and acetylsalicylic acid and these pharmaceuticals in isolated forminduce oxidative stress on Hyalella azteca. *Environmental Monitoring and Assessment*, **186**, 7259–7271, https://doi.org/10.1007/s10661-014-3925-0

Hurina I. V. (2014). Analiz suchasnykh pidkhodiv do klasyfikatsiyi medychnykh vidkhodiv v ukrayini. *Faces of Ukraine*, **4**(21), 51–54.

Khetan S. K. and Collins T. J. (2007). Human pharmaceuticals in the aquatic environment: a challenge to green chemistry. *Chemical Reviews*, **107**, 2319–2364, https://doi.org/10.1021/cr020441w

Kika L. S. (2020). Influence of antibiotics on the activity of activated sludge in the biological treatment of urban wastewater. Biotechnology of the XXI Century: Materials of the XIV All-Ukrainian Scientific-Practical Conference, May 20, 2020, Kyiv. KPI. Igor Sikorsky, Polytechnic Publishing House, Kyiv, 2020.

Korosteleva A. V. (2013). Methods of Wastewater Treatment From Phenols. Chemistry, Moscow.

Kozub P., Kozub S., Mozaffari N., Sihag P., Mozaffari N., Vambol S., Vambol V., Ramsawak N. and Khan N. A. (2021). Chapter 8: Treatment schemes – conventional and dedicated for PhACs treatment. In: Pharmaceutical Wastewater Treatment Technologies, N. A. Khan, S. Ahmed, V. Vambol and S. Vambol (eds), IWA Publishing, p. 181, https://doi.org/10.2166/9781789061338_0181

Maksyutina M. P. (1976). Analysis of Pharmaceuticals and Dosage Forms. Health, Kiev.

Novikov V., Sidorov Y. and Swede O. (2008). Trends in commercial biotechnology. *Visn. NAS of Ukraine*, **2**, 25–39.

Oliveira R., McDonough S. and Ladewig J. C. L. (2013). Effects of oxytetracycline and amoxicillin on development and biomarkers activities of zebrafish (Danio rerio). *Environmental Toxicology and Pharmacology*, **36**, 903–912, https://doi.org/10.1016/j.etap.2013.07.019

Paíga P., Santos L. H. M. L. M. and Ramos S. (2016). Presence of pharmaceuticals in the Lis river (Portugal): sources, fate and seasonal variation. *Science of the Total Environment*, **573**, 164–177, https://doi.org/10.1016/j.scitotenv.2016.08.089

Rede D., Santos L. H., Ramos S., Oliva-Teles F., Antão C., Sousa S. R. and Delerue-Matos C. (2019). Individual and mixture toxicity evaluation of three pharmaceuticals to the germination and growth of Lactuca sativa seeds. *Science of the Total Environment*, **673**, 102–109, https://doi.org/10.1016/j.scitotenv.2019.03.432

Strelnikov L. S., Strilets O. P. and Shcherbak O. V. (2006). Prospects and ways of development of production of biotechnological drugs in Ukraine. *Annals of Mechnikov Institute*, **4**, 3–8.

Tsalikov R. H. (2009). Assessment of Natural, Man-Made and Environmental Safety. MNHTU, Moscow.

Vambol S., Vambol V., Khan N. A., Mozaffari N. and Mozaffari N. (2021a). Algorithm for Elimination of the Consequences of the Oil-Polluted Water. In scientific works of the Third International Scientific and Practical Conference 'Modern Trends in the Development of Information Systems and Telecommunication Technologies', January 25–26, 2021, **Kyiv**, Ukraine. NUHT.

Vambol S., Vambol V., Mozaffari N., Mahmoud A. E. D., Ramsawak N., Mozaffari N., Ziarati P. and Khan N. A. (2021b). Chapter 2: Comprehensive insights into sources of pharmaceutical wastewater in the biotic systems. In: Pharmaceutical Wastewater Treatment Technologies, N. A. Khan, S. Ahmed, V. Vambol and S. Vambol (eds), IWA Publishing, p. 17, https://doi.org/10.2166/9781789061338_0017

Vambol V., Shulha A., Vambol S., Sihag P., Pavlykivskyi V., Mozaffari N., Mozaffari N., Khan N. A. and Sawicka B. (2021c). Legislative and criminal law aspects of water protection and prevention of pharmaceuticals accumulation in nature. In: Pharmaceutical Wastewater Treatment Technologies, N. A. Khan, S. Ahmed, V. Vambol and S. Vambol (eds), IWA Publishing, p. 113. https://doi.org/10.2166/9781789061338_0113

Vishnevetsky V. Y. and Vishnevetsky Y. M. (2004). Influence of Ecological Factors of Habitat on Public Health. Ecology, Taganrog.

doi: 10.2166/9781789062625_0149

Chapter 8

Sri Lanka National Hospital (NHSL) wastewater treatment

Tino Christen[1*] *Bogdana Marinova*[1], *Reto V. Schulthess*[1] *and Roland Metzler*[1]

[1]*HOLINGER AG, Liestal, Switzerland*
Corresponding author. christen_tino@hotmail.com

ABSTRACT

The National Hospital of Sri Lanka (NHSL) has to realise a (pre)treatment of its wastewater in order to renew the concession for discharge into the Colombo Municipal Council (CMC) wastewater system. The goal of this work was to find a feasible solution that would meet the framework and the given schedule. To achieve this goal, the project team visited the hospital to get detailed site information and gain a deeper understanding of the local challenges. Further meetings with the hospital's director to analyse the master plan and discussions with members of the Ministry of Health revealed the requirements that needed to be met. Basic data of wastewater quantity and quality was collected on site via water consumption data and measurements of wastewater streams. In the feasibility study, the solution with a membrane bioreactor (MBR) wastewater treatment plant within the NHSL for the joint treatment of all wastewater from the hospital proved to be the best option. Subsequently, the treatment plant was designed based on the collected data and the requirements of the hospital.

Keywords: Colombo, Sri Lanka, Hospital, NHSL, MBR, Wastewater treatment, WWTP

CONTENTS

8.1 GENERAL INTRODUCTION AND AIM OF THE PROJECT

The National Hospital of Sri Lanka (NHSL) is located in Colombo, Sri Lanka. Founded in 1864 as the General Hospital, it is the leading hospital of the country and is controlled by the central government. The hospital has 18 intensive care units and 21 operating theatres. The hospital treats over two million patients a year. Situated on a 36-acre site, it includes the Dental Institute, Maligawatte Kidney Hospital, Nurse's Training School, Post Basic Nurse's Training School, School of Eco Cardiograph, School of Physiotherapy, School of Radiography and the University of Colombo's Faculty of Medicine (Ministry of Health, 2021).

The hospital has grown organically over the years and its structure is still changing through renovation and replacement works. Its extent can be seen in Figure 8.1 and in full-size in Appendix 8.3. Hence, one can say that there was no master plan of development and as a result, the sewer system infrastructure was neglected. Wastewater and storm water lines are directly discharged into the Colombo Municipal Council (CMC) sewer system. Some of the major problems facing the hospital are sewer blockages, wastewater overflows and interconnections between the stormwater and wastewater system.

In the near future, Colombo will commission two municipal wastewater treatment plants (WWTPs). These WWTPs are designed by Chinese engineers. For the smooth-running operation of the centralised treatment plants, the Chinese have selected companies/factories with special wastewater characteristics and demanded from them a decentralised (pre-)treatment system. The NHSL is one of these chosen businesses.

The aim of this project is to treat the wastewater from the hospital to make it harmless for the centralised treatment plants. This means eliminating the toxic and hazardous substances. The deadline set by the Chinese for commissioning the decentralised treatment plant of the hospital is before the centralised WWTP goes into operation. Otherwise, the hospital is not allowed to discharge its wastewater to the CMC sewage system.

Western hospitals, rarely include a wastewater treatment system. Toxic and hazardous substances are separated or rendered harmless at the source and are

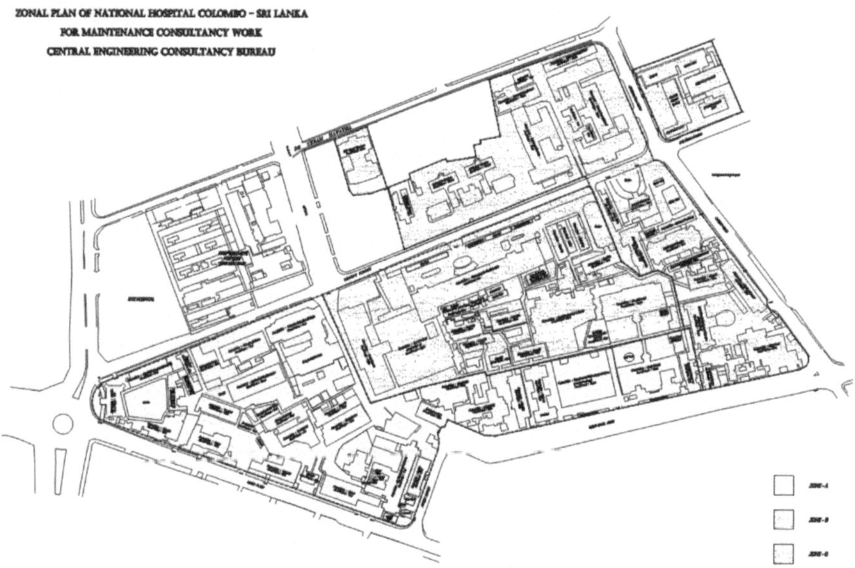

Figure 8.1 Site plan of the National Hospital of Sri Lanka.

not discharged to the sewer system. Therefore, the hospital wastewater does not differ greatly from municipal wastewater. Examples for treatment at the source are recovery of substances, neutralisation, cooling tank for radioactive radiation, incineration with solid waste or thermal treatment (autoclave). To achieve this, the infrastructure for the separate disposal of different types of waste and the proper education and training for employees (from cleaning staff to doctors) must be implemented.

With the NHSL's existing infrastructure, this does not seem possible in the foreseeable future. Retrofitting of the existing buildings is not cost effective. Nevertheless, a separation system should be designed in future buildings. This will result in a slow but continuous improvement of waste disposal as well as the wastewater system. But in order to meet the Chinese demand, a solution has to be found that can be implemented immediately.

8.2 FEASIBILITY STUDY

The framework conditions in the search for a suitable solution for the NHSL were as follows:

- Elimination of toxic and hazardous substances
- Elimination of pathogens and viruses
- Robust against shock loads
- Low space requirement
- Manageable maintenance
- Low CAPEX and OPEX (capital and operating expenditure)

Based on these framework conditions, the best solution was sought. Since the separation at source is not executable in the required timeframe, two feasible alternatives are either several packaged treatment plants at various locations within the hospital area or one large, decentralised plant for all wastewater.

Packaged treatment plants are compact wastewater treatment systems prefabricated by the manufacturer. They are usually built modularly in containers in order to be able to meet different treatment needs and wastewater quantities. There are many manufacturers of packaged plants on the market which promise different performances. For the NHSL, one solution would be to build such packaged treatment plants on every exit point of the NHSL to the CMC sewer containing problematic wastewater. However, the problem with this solution is that by distributing the wastewater to several small plants, the impact of shock loading of problematic wastewater on individual plants increases. This means that a shock load can cause lasting damage to the biology of a packaged plant which can take a long time to recover.

Furthermore, these packaged treatment plants are almost like 'Black Boxes'. The operation and maintenance of the plants can usually only be executed by the manufacturer and, therefore, the customer is very much dependent on them. Further, the monitoring of the operation is rather difficult.

With the second option, all wastewater from the entire hospital is collected and treated in a decentralised wastewater treatment plant. This results in good mixing of municipal wastewater and problematic substances from point discharges within the hospital. This wastewater can be processed without operational problems by one slightly bigger treatment plant. Further, the maintenance and operation (M&O) of one treatment plant is preferable to the M&O of several plants at various locations distributed over the entire hospital site.

From these considerations, one decentralised wastewater treatment plant was chosen to be the most feasible solution. The following chapters describe the dimensioning and the design of the decentralised wastewater treatment plant.

8.3 DESIGN DATA

8.3.1 Wastewater volume

The NHSL land extends about 36 acres at present. Its bed, staff and patient capacity and the corresponding wastewater production is listed in Table 8.1. There are no measurements of the wastewater quantity. Therefore, the wastewater production is assumed, based on values from the German guideline

Table 8.1 NHSL present wastewater production.

		Wastewater Production
Number of beds	3200	550 l/bed*d
Total number of staff	7500	150 l/PE*d
OPD patients	5000	150 l/PE*d

PE: population equivalent, OPD: outpatient department.

Table 8.2 Future wastewater production.

		Wastewater Production
Number of beds	5000	450 l/bed*d
Total number of staff	11 250	150 l/PE*d
OPD patients	7500	150 l/PE*d

DWA-M 775 (DWA, 2010) and verified with the water consumption data, which was available from the supply branch of NHSL.

The calculated total wastewater volume from the patients and staff is 3635 m^3/d.

The water consumption data was collected from March to August 2018 in cubic metres per month per building of the NHSL. From this data, the daily water consumption was derived. It is equal to 3800 m^3/d. This corresponds to the amount of wastewater. The difference may be due to irrigation of green areas where no wastewater is produced due to infiltration.

The hospital is planning to expand with a new OPD building to the northwest. Also, the replacement of old one-storey buildings with multilevel buildings inside the hospital area is planned, to increase the capacity. The expected growth for the next 20 years is assumed to be 50%. As a result, the expected capacity in 2040 is as shown in Table 8.2.

The wastewater production per bed is expected to decrease due to improved water management. The volumes are expected to approximate the experience values known from western hospitals. Therefore, the predicted wastewater production is 5000 m^3/d.

In Table 8.3, the hydraulic design of the wastewater treatment plant is shown.

The wastewater volume is subject to high uncertainties because there are no measurements. Additionally, the sewer network condition is not well documented. Therefore, it is possible that a substantial amount of wastewater is lost due to leakage in the network system. Further, it may be possible that some of wastewater pipes are interconnected with the stormwater system which would either increase or reduce the wastewater amount. These factors can result in overestimation of the hydraulic capacity of the treatment plant. But with the continuous renewal of the sewer system inside of the hospital, which is progressed along with the renewal of the hospital itself (according to the masterplan), the connection rate and therefore the percentage of treated wastewater will increase.

Table 8.3 Hydraulic dimensions of the treatment plant.

Daily wastewater production	5000	m^3/d
Q_{mean}	208	m^3/h
Q_{peak} (1.5 * Q_{mean})	313	m^3/h
Q_{mean}	58	l/s
Q_{peak} (1.5* Q_{mean})	87	l/s

8.3.2 Wastewater concentration

Wastewater concentrations and characteristics from literature could not be used directly, since the wastewater composition varied significantly from the different buildings inside the NHSL area. But the different sources of wastewater were analysed according to the literature and toxic and hazardous substances were evaluated (Vambol *et al.*, 2021a). Therefore, local partners analysed the wastewater at five different manhole locations. The locations were chosen to provide the best overview of the whole hospital area. The results are presented in Table 8.4.

The results show a similar wastewater characteristic as that known for domestic wastewater. The COD measurement from manhole number 300 shows very high values, but the BOD is comparable to domestic wastewater. It is possible that the COD consists mostly of refractory COD, which is hardly biodegradable. In general, the BOD/COD ratio is low. The PO_4 concentration is higher than that known for domestic values. A possible reason for this is that the hospital may use detergents with a high phosphorus content. If so, the detergents should be replaced with phosphorus-free agents. The pH is in the

Table 8.4 Wastewater analysis.

Sample Point (Manhole Number)		pH	BOD [mg/l]	COD [mg/l]	NH_4-N [mg/l]	PO_4 [mg/l]
244	Morning	6.8	62.9	300.0	20.0	3.8
	Midday	7.4	238.4	130.0	39.0	6.2
	Evening	6.8	68.7	200.0	18.0	7.6
		7.0	123.3	210.0	25.7	5.9
300	Morning	6.8	302.9	2300.0	32.0	5.4
	Midday	7.2	315.0	1190.0	69.0	42.0
	Evening	6.8	96.0	830.0	69.0	55.0
		6.9	238.0	1440.0	56.7	34.1
178	Morning	6.9	83.7	440.0	8.0	1.3
	Midday	7.2	96.0	800.0	38.0	6.0
	Evening	6.8	138.0	540.0	4.0	6.9
		7.0	105.9	593.3	16.7	4.7
10	Morning	8.9	73.7	740.0	17.0	1.2
	Midday	9.9	70.4	200.0	11.0	17.3
	Evening	7.6	81.9	990.0	34.0	21.0
		8.8	75.3	643.3	20.7	13.2
135	Morning	7.9	114.6	460.0	15.0	2.5
	Midday	7.8	92.4	950.0	89.0	26.8
	Evening	7.3	100.0	240.0	24.0	15.0
		7.7	102.3	550.0	42.7	14.8

The map showing the manhole numbers can be seen in Appendix 8.2.
BOD, biochemical oxygen demand; COD, chemical oxygen demand.

Table 8.5 Wastewater concentration and loading.

	Concentration [mg/l]	Load [kg/d]
BOD	150	750
COD	600	3000
NH_4-N	30	150
PO_4	10	50

range of domestic wastewater and therefore no special attention needs to be paid to that parameter.

For the design of the wastewater treatment plant, the following concentrations and wastewater load were chosen (Table 8.5).

The load is calculated by multiplying the concentration by the expected future daily wastewater volume of 5000 m^3/d.

8.3.3 Sri Lanka discharge standards

The discharge standards in Sri Lanka recommended by the Central Environmental Authority (CEA) are listed in Table 8.6.

The discharge standards in Sri Lanka are not very strict. Nevertheless, the goal of the treatment of the hospital wastewater is to eliminate pathogens, and toxic and hazardous substances. Hence, the design of the treatment plant is intended to achieve more stringent effluent concentrations.

8.4 WASTEWATER TREATMENT PLANT DESIGN

The wastewater treatment design is based on the documents DWA-A 131 (DWA, 2016) and DWA-A 227 (DWA, 2014). The electromechanical equipment was chosen in cooperation with a local supplier. The plant is designed to

Table 8.6 Sri Lanka discharge standards.

Discharge Standard Sri Lanka	Unit	Industrial Waste into Surface Water
TSS	mg/l	50
pH	–	6.0–8.5
BOD	mg/l	30
COD	mg/l	250
Phosphates	mg/l	5
N_{tot} Kjeldahl	mg/l	150
Ammonium	mg/l	50
Phenolic compounds	mg/l	1.0
Sulphides	mg/l	2.0
Fluorides	mg/l	2.0
Total residual chlorine	mg/l	1.0

TSS, total suspended solids; N_{tot} Kjeldahl, total Kjeldahl nitrogen.

accommodate the growth of the hospital over the next 20 years, because the infrastructure which is built has an operating life of 20 years. The wastewater load should not exceed the capacity of the treatment plant in this time horizon.

8.4.1 Wastewater ring collection line

To collect all wastewater leaving the hospital area, a ring collection line is planned. This collection line is located at the border of the hospital area and connects all exit points. Since the location of the treatment plant as not yet been determined, the requirement for pumping stations around the hospital is not clear. The collection line needs a minimal downward slope to guarantee a minimal flow velocity to ensure the transport of the wastewater without clogging the line. Hence, pumping stations may be needed. The topographic elevation of the ring collection line around the hospital area is shown in the map in Appendix 8.4.

8.4.2 Inlet pumping station

The inlet pumping station elevates the wastewater from the pumping pit to the inlet of the wastewater treatment plant. The pumping pit is located at the lowest point of the ring collection pipe to catch all wastewater from the hospital.

The pumping station consists of two wastewater pumps each with 90 l/s capacity. Therefore, each pump individually has the capacity to process the predicted Q_{mean}. In peak hours, both pumps will be working. This redundancy guarantees that during maintenance activities, the Q_{mean} flow can still be transported to the treatment plant.

Centrifugal pumps: 2×90 l/s

An overflow to the CMC sewer system prevents the pumping pit and ring collection from overflowing.

8.4.3 Mechanical pretreatment

The mechanical treatment consists of two identical treatment lines.

8.4.3.1 Fine and coarse screen

The biological treatment of this wastewater treatment plant is a membrane bioreactor (MBR). To prevent the membranes from clogging, the mechanical treatment should remove most suspended particles such as rags, fibres, hair, grit and so on. Therefore, two consecutive screens are installed. The first one is a coarse screen with 6 mm bar space and the second a fine screen with 1–2 mm bar space. These bar spaces were selected based on manufacturer recommendations. Each treatment line has a hydraulic capacity of 90 l/s.

The removed solids are transported to the screenings wash press for dewatering. The solids can be dewatered to 60% dry solids to substantially reduce transport and disposal costs. The dewatered solids are stored in containers, for ease of collection by lorry.

Coarse screen 6 mm: 2×
Fine screen 1–2 mm: 2×
Wash press: 1×

The fine and coarse screens and the screenings wash press are housed to prevent odour emissions.

8.4.3.2 Grit chamber

The grit chamber removes sand and grit from the wastewater. Through sedimentation, particles bigger than 0.1–0.2 mm can be removed. The settled sand and grit at the bottom will be removed periodically using a suction vehicle or an excavator.

The hydraulic residence time of the grit chamber is the key parameter for the design. Typical values are from 5 to 10 minutes.

The peak flow is 313 m^3/h. The grit chamber is again constructed in two identical lines each with a design flow of 160 m^3/h. During maintenance of one line, the wastewater can still be transported adequately, but the removal efficiency will be reduced. With a Q_{peak} of 320 m^3/h and hydraulic residence time of 10 minutes the following minimal total volume of both grit chambers is necessary: 52 m^3.

The design of each grit chamber is defined as follows:

Length grit chamber:	8 m
Width grit chamber:	1.5 m
Depth grit chamber:	2.3 m
Total volume (both chambers):	$2 \times 27.6\ m^3 = 55\ m^3$

The aim of the grit chamber is to remove sand particles from the wastewater but to let the suspended organic particles pass.

No primary clarifier is planned since the handling of the primary sludge is difficult and the organic particle removal can also be achieved in the bioreactor. As a result, the biology must be built correspondingly larger. But on the other hand, the large surface area of a primary clarifier can be spared.

8.4.4 Biological treatment: membrane bioreactor

8.4.4.1 Variant decision

As already mentioned, the determining factors for the evaluation of the treatment system are:

- Elimination of toxic and hazardous substances
- Elimination of pathogens and viruses
- Robust against shock loads
- Low space requirement
- Manageable maintenance
- Low CAPEX and OPEX

Before discussing a wastewater treatment plant to cover all wastewater of the hospital, the possibilities of source separation or treatment of only highly polluted streams were considered. However, the installation of a complete separation system inside of every building and new sewer lines between the buildings was considered to be impossible. Also, the sources of these substances are scattered over the whole area, therefore many packaged treatment plants would be needed. For the proper operation of a source separation system, education of the staff and proper infrastructure are the most important factors. The education of the staff is a long-term process and therefore not feasible for a prompt successful solution. Nevertheless, the correct disposal of hazardous and toxic substances and medicaments should be an important long-term goal.

Therefore, the solution of collecting all wastewater exiting the hospital area and treating it all together, was considered to be the most feasible and cost-effective variant. With the chosen process, the MBR system, the toxic and hazardous substances as well as bacteria and pathogens can be removed from the wastewater. Hence, the sources, quantity and exact composition do not need to be known in detail (which saves substantial effort for laboratory and on-site analytics). Another advantage of collecting all the wastewater together is that shock loads of problematic substances are diluted and therefore the possibility to impair the treatment is low. The wastewater and substances will all be transported through the existing sewer network, collected using the projected collection line and treated at the decentralised treatment plant.

8.4.4.2 Process description

The MBR is based on the conventional activated sludge process, but the separation of microorganisms is performed by filtration with membranes (Vambol *et al.*, 2021b).

The MBR system works on the principle of solid separation through the concept of using flat sheet or hollow fibre submerged membrane modules in the bioreactors. The membranes in an MBR system are made from polymeric organics (polyvinylidene difluoride (PVDF), polyethylene (PE) or polyethersulfone (PES)) and assembled into units (modules, cassettes, stacks) with high packing density. Raw wastewater pretreatment is important to sustain stable MBR performance and fine screening is essential for the operation.

The use of MBRs in municipal wastewater treatment has grown widely in recent decades. The MBR technology combines conventional activated sludge treatment with low pressure membrane filtration, thus eliminating the need for a secondary clarifier or polishing filters. The membrane separation process provides a physical barrier to contain microorganisms, pathogens and some viruses and assures consistent high-quality water.

The NHSL management stated that there is limited space available and therefore a compact design is necessary. The MBR has the most compact design of all the activated sludge processes since the activated sludge tank can operate with a mixed liquor suspended solids (MLSS) concentration of 10 g/l, which is three times higher than with conventional treatment systems. This means the biological tank can be built substantially smaller. Further, the solid separation through the membranes makes the land-intensive secondary clarifier (settler) unnecessary (Figure 8.2).

With the high MLSS concentration, the MBR can operate at long sludge retention times (SRTs). This has the effect, that peak loadings of toxic and hazardous liquids do not overstress the biological system. Further, through long SRTs, specialised microorganisms can grow and degrade trace substances

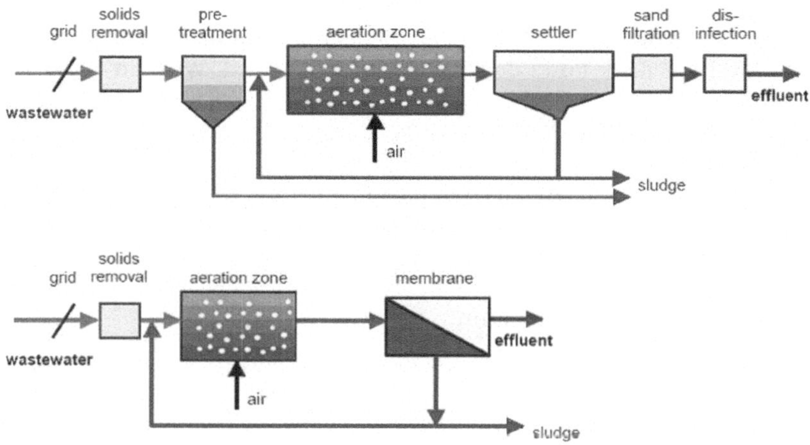

Figure 8.2 Comparison of the treatment train of conventional activated sludge system (top) and the MBR system (bottom) without primary clarifier.

which are untreated in conventional systems. Pathogens and most viruses are killed in the system and retained by the membranes.

Further advantages of the MBR:

- Capable of meeting high water quality standards for effluent.
- Membrane modules are back-flushable.
- Requires chemical cleaning only once in 3 to 6 months.
- Less excess sludge than conventional system.
- Compact design: footprint 75% smaller than conventional system.
- No secondary clarifier, virtually no effluent suspended solids.
- High MLSS can be maintained.
- Easily automated and instrumented to measure performance.
- It allows systems to be remotely operated and monitored, thus significantly reducing operator attendance.
- Membrane durability could be significantly increased, and the production costs decreased thanks to R&D in recent years.

Disadvantages of the MBR:

- Limited tolerance for abrasive and stringy materials, such as grit, hair and fibrous material→adequate mechanical treatment required.
- Accumulation of solids and sludge between membrane fibres and plates can clog/damage the membrane tube openings→prevention by using air scouring.
- Membrane fouling over time.
- Higher energy consumption to overcome trans-membrane resistance.
- High aeration requirements due to high MLSS concentration.
- Broken membranes result in low effluent quality→early detection needed.

8.4.4.3 MBR design

To design the MBR system to be as cost effective as possible, an equalisation tank to level the peak flows during the day is planned. With an equalisation tank, the MBR can be designed for the $Q_{design} = Q_{mean}$ flow and not for the Q_{peak}, which results in a smaller infrastructure.

Q_{design} MBR:	210 m^3/h (Design flow of the MBR)
Q_{peak}:	313 m^3/h (Flow during peak hours)
Equalisation tank:	320 m^3 (LxWxH: $8 \times 8 \times 5$ m)
Retention time:	3 h (320 $m^3/(Q_{peak} - Q_{design})$) $\rightarrow$ If the inflow is equal to Q_{peak} for longer than 3 h, then overflow to CMC sewer

Stirrer to prevent sedimentation

The biological tank contains activated sludge which degrades the organic substances. The microorganisms need an oxygen-rich environment; therefore, the tank must be aerated by aeration plates and blowers. The long solids residence time leads to a part stabilisation of the sludge. This means that the excess sludge has a lower biological reactivity and therefore less odour nuisance and a lower possibility of harmful contamination. The whole MBR system consists of two identical treatment lines. The excess sludge is withdrawn directly from the biological tank. The design of the biological tank is defined as follows:

Number of tanks:	2
MLSS concentration:	10 g/l
Solids residence time:	20 d (SRT)
Aeration:	Turbo-blower, aeration plates
Volume per tank:	750 m^3
Length:	20 m
Width:	7.5 m
Depth:	5 m
Total volume:	1500 m^3
Total surface:	300 m^2

After the biological treatment, the wastewater flows into the membrane filtration tanks. The membranes are installed in trains. The system consists of four trains, two trains after each biological tank. The filtration is designed in such a way that three trains can guarantee the Q_{design} flow. To prevent the membranes from clogging, air scouring is applied. This means that the membranes are periodically strongly aerated to remove deposition on the surface. Permeate pumps suck the wastewater through the membrane separating the treated water from the solids. The permeate is stored in a tank for backpulsing of the membrane. CIP (clean in place) is the chemical cleaning of the membranes without removing them from the tank. The dimension of the membrane filtration tanks is:

Number of trains:	4
Design temperature:	25°C
Filtration tank depth:	4 m
Tank surface per train:	12 m^2

Total surface:	48 m^2
Total tank volume:	192 m^3
Membrane surface:	3000 m^2 per train
Total mem. surface:	12 000 m^2
Membrane flux mean:	23 lmh
Permeate tank & CIP:	Membrane manufacturer information
Measurement:	Dissolved O_2, flow, pressure, airflow, water level, turbidity

Phosphorus removal is achieved by chemical precipitation. Metal salts are added into the return activated sludge. To achieve maximal precipitation, the dosed metal salts and the wastewater must be completely mixed.

The operating building contains the installations for the membrane peripherals and the control system of the treatment plant. It hosts the blower station, the return activated sludge system, the permeate pumps and the process monitoring system.

Building surface: 300 m^2

The **detailed design** of the treatment plant needs to be specified in close cooperation with the membrane manufacturer and supplier, because the requirements for mechanical treatment, MLSS concentration, membrane cleaning (hydraulic, chemical), air scouring and so on. vary greatly among different membrane manufacturers.

The effluent of the MBR can be discharged to a surface water and does not have to be discharged into the CMC sewer.

8.4.5 Optional disinfection: UV-light

For disinfection of the wastewater after the biological treatment, a UV-light disinfection can be considered. The effluent from the MBR reactor has very low suspended solids concentration because of the membrane liquid/solid separation, therefore the disinfection with UV-light would be efficient.

8.4.6 Sludge treatment

The MBR reactor will produce a daily amount of excess sludge of up to 75 m^3/d with a solids concentration of 10 g/l, which is approximately 750 kg TSS/d. This sludge can be disposed of according to the following sludge treatment train:

The excess sludge is first thickened with a rotary disc thickener up to 6% dry solids (DS) and subsequently dewatered with a decanter up to 20% DS and stored for transport. The sludge should be removed regularly to avoid odours and hygiene problems.

Since no further on-site treatment to stabilise the sludge is planned, the sludge should either be transported to the sludge treatment of the central wastewater treatment plant of Colombo or directly to an incineration plant for thermal oxidation. Direct application on agricultural land as fertiliser is not recommended.

8.4.7 Total footprint

The treatment line for the wastewater treatment plant should fit on an area of 40 m × 40 m (Figure 8.3). See the full-size site plan in Appendix 8.1.

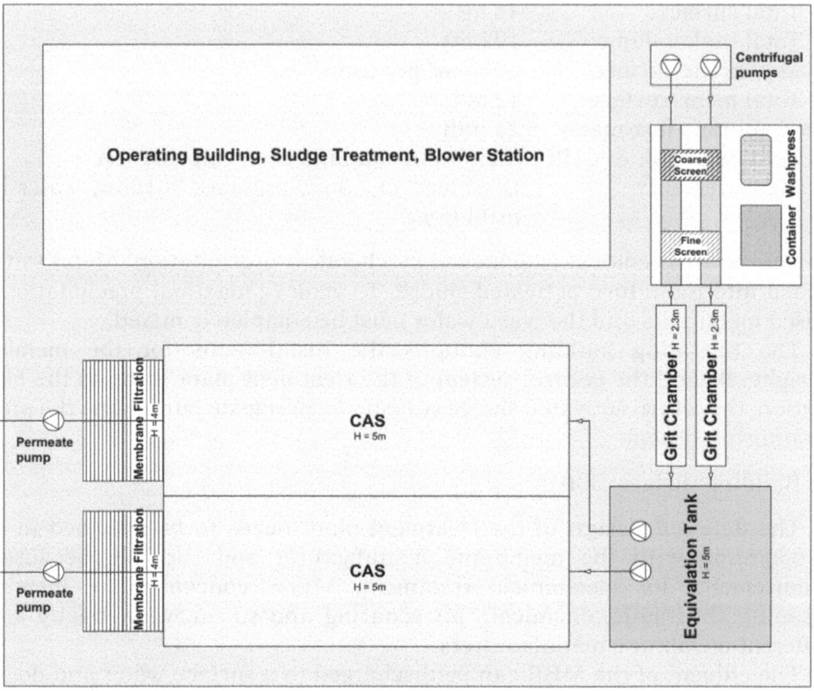

Figure 8.3 Layout concept of the decentralised MBR treatment plant.

8.5 CONCLUSION

The national hospital has to provide a (pre)treatment for its wastewater. The wastewater quantity was defined according to data from the literature and verified with water consumption data collected from the hospital. An analysis of the wastewater characteristics was conducted. The treatment plant should be designed to meet the capacity of hospital growth according to the masterplan for the next 20 years. The most feasible solution is to collect all wastewater from the hospital and treat it with a membrane bioreactor plant. With this option, toxic and hazardous substances, pathogens and viruses can be eliminated, and the design is robust against shock loads. Furthermore, the MBR treatment system has a very low space requirement, which is in highly favoured by the hospital's administration.

REFERENCES

DWA. (2010). Merkblatt DWA-M 775: Abwasser aus Krankenhäusern und Anderen Medizinischen Einrichtungen. Deutsche Vereinigung für Wasserwirtschaft, Abwasser und Abfall, Hennef, Deutschland.

DWA. (2014). Merkblatt DWA-M 227: Membran-Bioreaktor-Verfahren (MBR-Verfahren). Deutsche Vereinigung für Wasserwirtschaft, Abwasser und Abfall, Hennef, Deutschland.

DWA. (2016). Arbeitsblatt DWA-A 131: Bemessung von Einstufigen Belebungsanlagen. Deutsche Vereinigung für Wasserwirtschaft, Abwasser und Abfall, Hennef, Deutschland.

Ministry of Health. (2021). Wikipedia. Von Wikipedia: https://en.wikipedia.org/wiki/National_Hospital_of_Sri_Lanka abgerufen (last accessed 23 August 2021)

Vambol S., Vambol V., Mozaffari N., Mahmoud A. E., Ramsawak N., Mozaffari N. and Ziarati P. (2021a). Comprehensive insights into sources of pharmaceutical wastewater in the biotic systems. In: Pharmaceutical Wastewater Treatment Technologies: Concepts and Implementation Strategies, N. A. Khan, S. Ahmed, V. Vambol and S. Vambol (eds), IWA Publishing, p. 30–71. https://doi.org/10.2166/9781789061338_0017

Vambol S., Vambol V., Mozaffari N., Mahmoud A. E., Ramsawak N., Mozaffari N. and Ziarati P. (2021b). Chapter 8: Treatment schemes – conventional and dedicated for PhACs treatment. In: Pharmaceutical Wastewater Treatment Technologies: Concepts and Implementation Strategies, N. A. Khan, S. Ahmed, V. Vambol and S. Vambol (eds), IWA Publishing, pp. 237–291. https://doi.org/10.2166/9781789061338_0181

APPENDICES

Appendix 8.1: WWTP concept layout

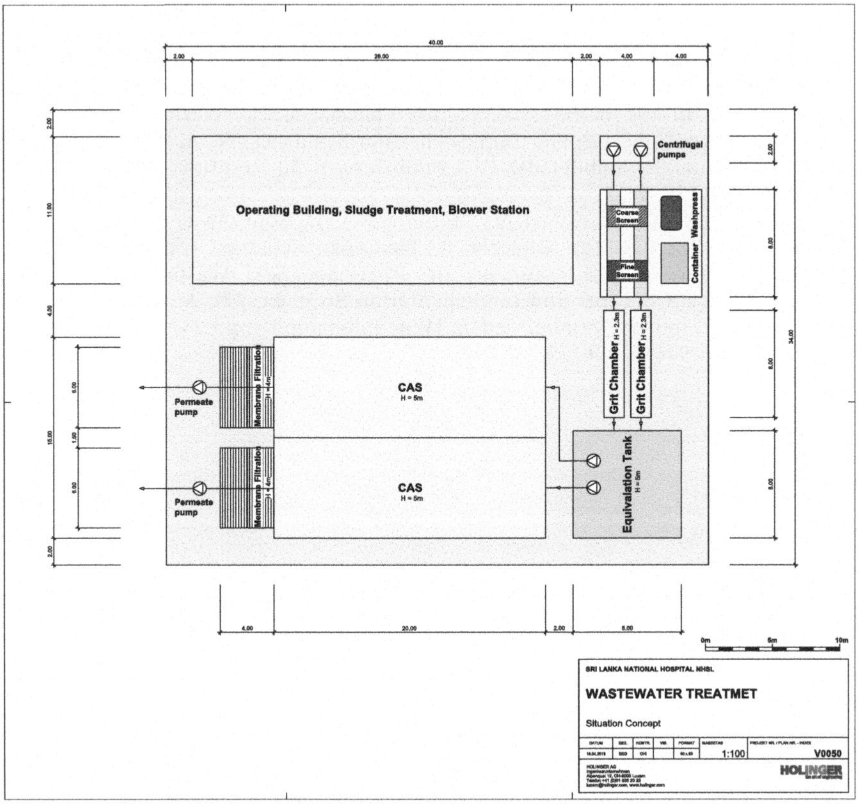

Appendix 8.2: Manhole location map

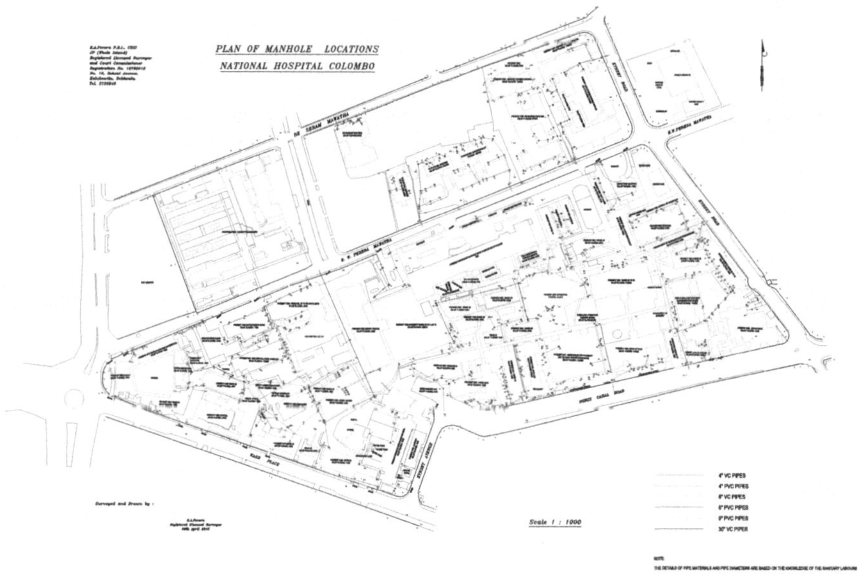

Appendix 8.3: NHSL layout plan

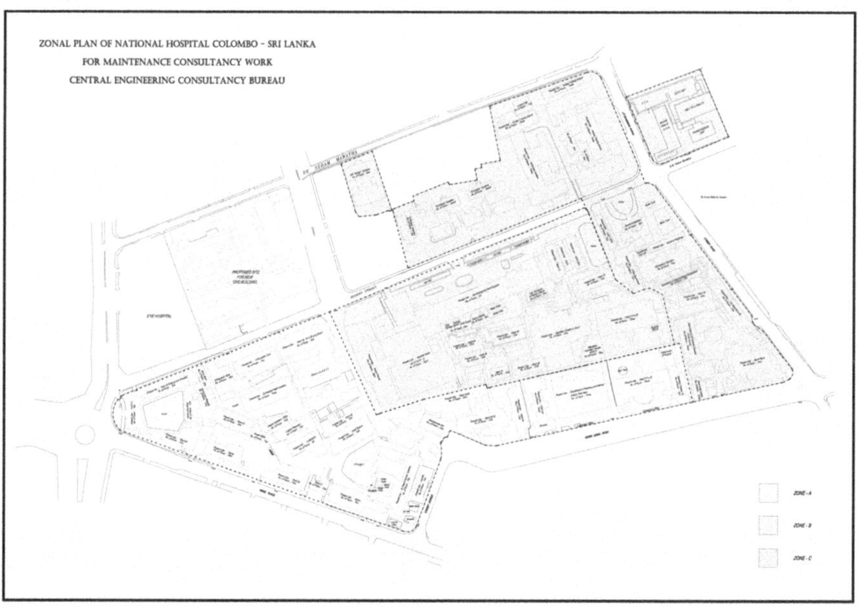

Appendix 8.4: Topographical plan of outer boundary of NHSL showing manholes

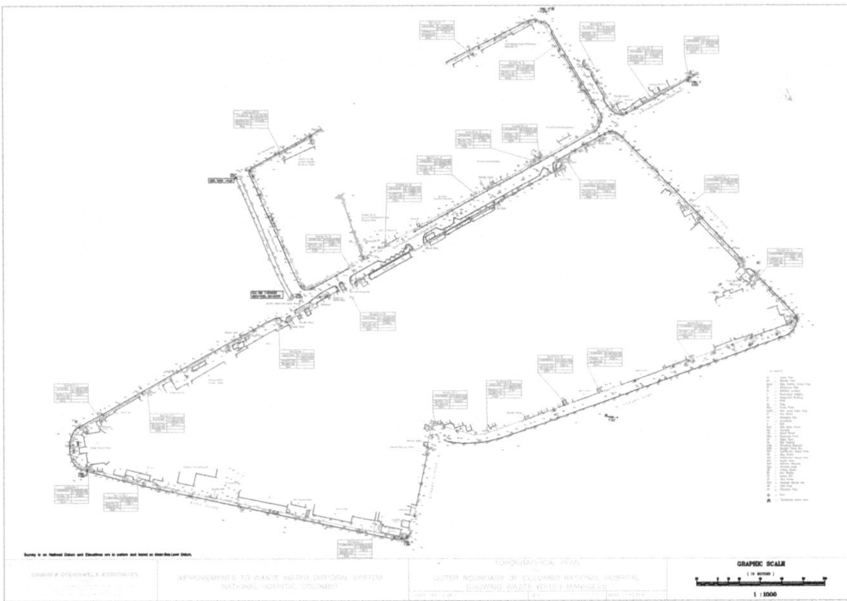

IWA PUBLISHING

doi: 10.2166/9781789062625_0167

Chapter 9

Hospital wastewater treatment scenario in India

Rishi Rathi

Environmental Science and Engineering Department, Indian Institute of Technology, Powai, Mumbai 400076, India
Corresponding author: rishirathi.iitb@gmail.com

ABSTRACT

Freshwater resources across the world are rapidly depleting. In order to keep up with the rapidly increasing population, urbanization and increasing standard of living, municipalities need to do more to reduce water wastage and reuse the wastewater after due treatment. India is a developing country and the infrastructure for proper wastewater treatment is still under-developed in several parts of the country. Hospital wastewater, when discharged without prior treatment, can be a potential risk to the environment and public health and therefore its treatment is essential. In India, however, the majority of hospital wastewater is discharged directly into the municipal sewer. Although the country has drafted legislation (Biomedical Waste Management and Handling Rules 1998), which was revised in 2016, most of the medical waste, including wastewater, is often dumped without prior treatment. For the small percentage of hospital wastewater that gets treated, the treatment procedure varies from state to state. For instance, in states like Punjab and Karnataka, the hospital wastewater is first disinfected on-site and then sent to the municipal sewers, whereas, in Maharashtra, the wastewater is first neutralized in waterproof buckets and then sent to the drains. In the course of this chapter, we analyse the existing data on hospital wastewater generation in India and study its composition. We then take a look at the effects of conventional disposal mechanisms on public health and the environment. This is followed by a discussion on the treatment processes being followed across different regions of the country and assessment of their effectiveness. Apart from this, we also look at the regulations and the legal clauses associated with hospital wastewater treatment. After studying all the aforementioned topics, we present a set of recommendations that can be implemented to reduce the environmental and public health impact of hospital wastewater discharge in the country.

Keywords: Diseases, environmental impact, hospital wastewater, pharmaceuticals, sewage

Editor(s): Nadeem A Khan, Viola Vambol, Sergij Vambol, Nastaran Mozaffari and Niloofar Mozaffari

CONTENTS

9.1 INTRODUCTION

Hospitals play a significant role in ensuring the well-being of the modern day human society. They treat diseases and save lives and through medical research advancements, they also protect mankind for the future. Different units at a hospital have varying water requirements to carry out their respective functions. In these processes, the water often gets contaminated and large amounts of wastewater are generated. The quantity of the wastewater as well as its chemical composition heavily depend on the purpose for which the water was used, the size of the ward/unit, the number of patients being treated, the service provided to the particular ward (canteen, laundry, air conditioning, medical purposes), the treatment practices being followed by the ward and the hospital, and the general policies and norms of the regions, that are binding on these institutions (Chandra, 1999).

The treatment of wastewater collected from hospitals is a significant challenge faced by municipalities across the world. Especially, in the past year, with the COVID-19 pandemic and the induced lockdowns, the stress on medical services has increased drastically accompanied by an increase in the hospital wastewater generation rates. This type of water needs special treatment because of its unique chemical and biological constitution (Kataki *et al.*, 2021). Wastewaters that are generated at hospitals typically contain pollutants such as residues of pharmaceutical matter, pathogenic organisms, chemicals that are used in medical treatment, radionuclides and other polluting matter. Most of these constituents are unique to hospitals and are not usually found in the sewage collected from the other parts of the city. Moreover, some of the constituents can be harmful to the treatment system and the ecology at large and need to be eliminated right at the source of their generation.

India, although a fast growing economy, is still a developing country. In spite of having strong legislation with regard to wastewater handling, the country still struggles with enforcement. Wastewater handling infrastructure is often expensive and thus very few hospitals (mostly private) can afford to integrate these technologies. In the majority of the hospitals, the wastewater is discharged into the municipal sewage system without any type of prior treatment (Mubedi *et al.*, 2013; Vambol *et al.*, 2021b). This kind of wastewater, upon getting mixed directly with wastewater of a common wastewater treatment plant, poses a major threat to the environment because these common treatment plants are not designed to treat contaminants from hospital wastewaters. A report by Sharma *et al.* (2015) established that this type of harmful treatment practice has led to the endangerment of some species of white-romped sharks and has enabled the genetic transformation of male fishes to female, because of the high concentration of contraceptive hormones and pain killers in marine systems such as rivers and streams.

In the course of this chapter, we will start off by analysing the existing data on hospital wastewater generation in India and study its composition. We will then study the effects of conventional disposal mechanisms on the overall environment and public health. We will then look into the treatment procedures being followed across the different regions of the country through three case studies – New Delhi, Central India and Lucknow (in Uttar Pradesh) – and assess their effectiveness. In addition to this, we will also study the legislative clauses associated with hospital wastewater treatment in India, including the Biomedical Waste Management Rules of 2016, mandated by the Government of India and the Centre for Pollution Control Board of India. After studying all the aforementioned topics, we will discuss a set of recommendations that can be implemented in order to reduce the environmental and public health impact of hospital wastewater discharge in the country through deployment of efficient yet cost-effective technologies and more stringent enforcement of government guidelines and standards. This chapter is aimed at giving an in-depth perspective of what hospital wastewater management looks like in a developing country like India. While the management of wastewater may vary based on parameters such as the location, the funding of the hospital, the size of the hospital and several other factors, the case studies discussed in this chapter are representative of the general wastewater management practices across the country. The recommendations discussed at the end may be applied in the case of several other developing countries too, where wastewater management is under developed.

9.2 HOSPITAL WASTEWATER GENERATION SCENARIO IN INDIA

9.2.1 Water supply to hospitals

Water is one of the most critical utilities in any hospital and therefore water supply to hospitals has to be properly planned and executed. In addition to this, timely maintenance is also necessary to avoid any unforeseen problems. According to the CPHEEO (Central Public Health and Environmental Engineering Organisation) Manual, the following quantities of water have been prescribed for Indian hospitals (CPHEEO, 1993):

(1) Number of beds exceeding 100 = 450 litres per head per day
(2) Nurses/staff quarters = 135 litres per head per day
(3) Minimum hot water requirements = 45 litres per head per day

The Bureau of Indian Standards has also estimated the water consumption in hospitals in India. According to their estimate, for hospitals with more than 100 beds, the average water consumption is around 450 litre/head/day (which roughly equates to 164 250 litres of water/head/year). On the other hand, for hospitals with less than 100 beds, the average water consumption is around 340 litre/head/day (which equates to 124 100 litres of water/head/year) (Bureau of Indian Standards, 2010).

As far as the quality of the incoming water is concerned, water supply has been mandated to meet drinking water quality criteria prescribed by IS 10500:2012. This is because it is this water which later enters the distribution system for use in important functions such as carrying out dialysis or performing surgeries or use in any other any section that requires highly stringent water quality (Balwani & Nagarnaik, 2017).

Some of the key use cases of water at a hospital include the following:

(1) Sanitation – this includes water that is used for cleaning and hygiene purposes. It consists of the water coming from taps, showers and toilets.
(2) HVAC systems – HVAC stands for Heating, Ventilation, and Air Conditioning systems. These are important to for critical care and refrigeration systems in hospitals.
(3) Medical processes – several medical processes such as hydrotherapy, dialysis, first aid, emergency operations, radiology and surgery require clean water.
(4) Canteens and kitchens – this includes water that is used in kitchens and canteens for preparing food for the hospital residents and patients.
(5) Laundry – several residential hospitals provide laundry services where clothes are cleaned to maintain the hospital hygiene.

The above-mentioned processes consume more than 90% of the total water used at a hospital (Watersense, 2012). All of these activities pollute water to varying extents thereby modifying the water's chemical composition significantly.

9.2.2 Constitution of hospital wastewater

The wastewater generated from hospitals contains several types of chemicals, pharmaceuticals and pathogens in varying concentrations. Their use cases at hospitals vary from laboratory and research waste to medical treatment purposes. The wastewater can also include excreta from patients. The major constituents of wastewater are drugs and their metabolites such as antibiotics; radionuclides; organic matters; metals; solvents; products of disinfection and sterilization; radioactive bio-markers; special detergents for endoscopes and other medical instruments; and iodinated contrast media. Metals are occasionally present in the form of preservatives used in diagnostic agents. Examples of these metals include iodinated X-ray contrast media, mercury, platinum and rare earth elements (such as gadolinium, indium, osmium) (Kumari *et al.*, 2020).

Although the overall composition is very complex and depends on the nature of services offered by the ward and the hospital, some of the most common constituents of hospital wastewater include the following:

(1) Pharmaceuticals – Pharmaceuticals are essentially medical drugs that can relieve the patients of illnesses and diseases through various processes (Khan *et al.*, 2021). Pharmaceuticals can further be characterized as follows:
   (a) Antibiotics: Antibiotics include the drugs that are extracted from microorganisms (such as certain types of fungi, bacteria, and actinomycetes) or other organisms. The most commonly used antibiotics include, sulfamethoxazole (sulfonamide), tetracycline, clarithromycin (macrolide), streptomycin (aminoglycoside), azithromycin, amoxicillin (β-lactam) and ciprofloxacin (fluoroquinolone) (Huang *et al.*, 2019). India has been identified as the world's largest consumer of antibiotics. The most commonly used drugs are the β-lactam antibiotics (Boeckel *et al.*, 2014).
   (b) Radiation and X-ray compounds: Radioactive materials can be defined as naturally occurring or man-made elements that are unstable and radioactive in nature. These can include α, β, and γ radioisotopes, and in the medical field they are typically used in many physical examinations of internal organs and tissues. Radionuclides and their isotopes are also used in examining and treating cancer patients. This is achieved through a process where the cancer cells are either shrunk in size or killed (Lee *et al.*, 2018).
   (c) Persistent compounds: Various types of psychoactive drugs are known to be of ubiquitous nature and tend to remain persistent in aqueous environments. These compounds are basically very slow to degrade in the environment. This normally results in the accumulation of these compounds and therefore, detrimental effects on the environment and public health, if the compound is toxic in nature.
   (d) Hormones: Hormones are widely used in agriculture and livestock breeding as well as clinical treatment procedures.

(2) Pathogens – In addition to pharmaceutical chemicals, hospital wastewater can also contain large amounts of pathogens such as bacteria, protozoa, viruses and certain types of parasitic worms. Because of the high concentrations of antibiotics in hospital wastewater, it is also very common for multidrug-resistant bacteria (MDRB) to be present in the wastewater. The most commonly found MDRB include extended spectrum beta-lactamase (ESBL)-producing organisms such as *Pseudomonas aeruginosa*, vancomycin-resistant *enterococci*, and *Escherichia coli*. These bacteria, viruses, and parasitic worm eggs are resistant to various environmental conditions and can therefore survive for long durations inside the sewage. These pathogens get carried along with the other wastewater, eventually polluting freshwater resources by entering rivers and streams. The crops and vegetables can become contaminated when the treated effluent from the wastewater treatment plant or river water is used for irrigation purposes. Eventually, these pathogens can also reach humans through the food chain and infect them. This kind of a scenario could even lead to a massive outbreak of pathogenic infections (Yang *et al.*, 2018).

9.2.3 Effects of hospital effluents

Some of the aforementioned contaminants are toxic, non-biodegradable or infectious in nature. These pollutants, even at very low concentrations, have been known to have an adverse toxic effect on humans and marine organisms and are hence referred to as biologically active substances (Desbiolles *et al.*, 2018). These wastewaters also contain pathogenic microorganisms such as bacteria, viruses, protozoans, fungi, and helminths which have the potential to induce quick organism adaptation to these varying conditions and wastewater compositions by the process of genome rearrangement. This genetic exchange also builds up a resistance trait in microorganisms towards the particular chemicals, thereby rendering them ineffective (Jolibois & Guerbet 2006). Primarily, the effects of these contaminants can be categorized as follows:

(1) Antibiotic resistance: For more than 70 years now, antibiotics have been used in the field of medicine and agriculture for preventing infections. Most of the antibiotics can be utilized 100% by the host organism and through urine and faeces, they enter the environment. These antibiotics lead to the development of ARGs (antibiotic resistance genes), which are basically genes that make the organism resistant to the particular antibiotic by enabling them to adapt themselves to the changed environment (Hoek *et al.*, 2021). This is achieved through various mechanisms:
   (a) Target bypass – through this mechanism, the antibiotic is unable to reach its target enzyme because of either mutations or absence of enzyme gene in the organism.
   (b) Efflux pump – in this case, the intracellular concentration of antibiotics gets reduced because of modifications in the structure of the cell membrane.
   (c) Antibiotic inactivation – here, the antibiotic molecules are deactivated directly.
   (d) Target modification – in this mechanism, the action sites for the respective antibiotic are modified.
   (e) Because of the overuse of antibiotics, they have become a very common contaminant in surface waters. In some areas, they have even been reported in groundwater samples. Apart from the chemical contamination, the development of antibiotic resistance among microorganisms is a very severe potential hazard. Humans, plants and animals come in contact with microorganisms through various means and therefore ARGs can restrict our ability to prevent future infections. Through sewage and agricultural runoff, the antibiotics reach water bodies, where they can lead to the growth of ARGs in the surrounding ecosystems. These ARGs then lead to the modification of bacterial species present in the ecosystems. Through the uptake of soil nutrients and water, these ARGs can then be transported directly to the nearby living organisms such as fishes, plants and animals. As they proceed through the food chain, these ARGs can also reach larger animals, including human beings (Vambol *et al.*, 2021a).
   (f) The development of ARGs can be harmful because when infected, the treatment procedures may fail. This might call for the use of

stronger and potentially harmful treatment dosages. As stronger antibiotics are applied in the treatment procedure, more powerful antibiotic resistance bacteria will be created. Then the treatment becomes much more difficult thereafter (Wang, 2018).

(2) Pathogens and pharmaceuticals in the environment: Hospitals are some of the biggest contributors to the introduction of pathogens and pharmaceuticals in the environment through their activities and waste and wastewater disposal. Through these means, these compounds and pathogens also find their way into the groundwater of the disposal sites and surface water such as lakes and rivers. Once in these water bodies, the contaminants can easily enter the food chain through plants and aquatic animals. Since the rate of metabolism of antibiotics is also very low, the majority of them re-enter the sewerage and contaminate water bodies further. In India, wastewater treatment mainly consists of these steps – screening, gritting, primary sedimentation, biological treatment, and in some cases, secondary sedimentation. This level of treatment is often not sufficient to remove antibiotics and other pharmaceuticals from wastewater (Al Aukidy *et al.*, 2014).

Hospitals also generate some amounts of radioactive waste materials. These materials are largely used in treatment procedures and diagnosis of patients who are injecting or taking radioactive isotopes. Several hospitals in India do not have the capacity to dispose of these wastes properly and the majority of it is allowed to enter the sewerage. These radioactive waste are known to have a damaging effect on marine ecologies.

Pathogens that are collected from hospitals also enter the sewage systems through wastewater routes. Because of leakages in pipelines and poor wastewater infrastructure, these pathogens often enter the soil and leach into groundwater reserves too. Therefore, pathogen pollution of the groundwater is a realistic possibility. Moreover, because of the lack of well-designed wastewater treatment facilities in most municipalities in India, the pathogens can easily enter surface water bodies such as rivers, lakes, and reservoirs through the effluents that are discharged into them from treatment plants. These bodies hence suffer from excess pollution when the effluents are discharged into them. Many pathogens, such as bacteria and viruses, are known to accumulate inside the bodies of higher organisms. When humans interact with these higher organisms (through food), the pathogens tend to be a major threat to public health. They cause digestive and intestinal disease in humans and animals. For instance, *Salmonella* paratyphi and *Salmonella* typhi are known to cause paratyphoid and typhoid, respectively (Ooi *et al.*, 2018).

The close contact of humans with pathogenic organisms can be the cause of several diseases. The improper discharge of poorly treated hospital wastewaters is causing significant bacterial and physicochemical pollution in the environment and this problem is becoming more and more serious, as freshwater sources are depleting. Pathogens like *Vibrio* and *Salmonella*, which are damaging for human bodies, have also been identified as the major causing agents of epidemics in several communities across the globe.

9.3 CASE STUDIES

9.3.1 New Delhi

Lamba *et al.* (2017), studied 12 hospital samples of wastewater and 5 sewage drains across the city of New Delhi over 2 seasons. It was observed that the local wastewater treatment facility designed for primarily domestic wastewater, is more often than not, inadequate, and the treatment doesn't assure clean, safe water. For instance, it has been observed that only approximately 40% of the total wastewater in New Delhi is actually being treated. The hospital effluents contaminate the community sewers and contribute to public health concerns such as antibiotic resistance and pathogenic outbreaks. This is especially concerning because the hospitals in Delhi use high loads of antibiotics in treatment, are often overcrowded, and an efficient wastewater treatment set-up is often not available in these poor hospitals. The poor and the marginalized who live along drains in their unorganized settlement areas, are highly prone to becoming infected by this wastewater that is improperly discharged. Because of the above-mentioned reasons, antibiotic resistance infections are becoming more and more common, which in turn leads to higher per patient treatment costs, prolonged morbidity among patients, and increased death rates because of the rapid transmission of organisms with multidrug-resistance (MDR). Examples of such microorganisms include *Mycobacterium tuberculosis, Vibrio cholera, Streptococcus pneumoniae, Salmonella sp.* and *Shigella.*

In general, it has been well established that the wastewaters originating at hospitals are a key source of antibiotic resistance. However, information and research about the impact of hospital wastewater and waste on the environment, the excess of antibiotic resistant bacteria and antibiotic resistant genes entering water bodies and surrounding environments, and the difference between the hospital wastewater discharge quantities and the background discharge quantities in community sewage is still lacking (Gupta *et al.*, 2009). In order to answer these questions, ARB (antibiotic-resistant bacteria)/ARGs from a total of 12 hospital different wastewater samples and 5 sewage drains in New Delhi were tested and studied for 2 different seasons, in order to study the seasonal variation too.

Triplicates of wastewater samples from outfalls of 12 hospitals in New Delhi were collected during February–March (winter season) and May–June (summer season) in 2014. Hospitals were selected on the basis of various criteria such as the bed capacity of the hospital, whether it is privately or publicly funded, the location of the hospital and mode of treatment practised at the hospital (unani or allopathic). The majority of the surveyed hospitals in the study were government funded public hospitals that used both allopathic as well as conventional procedures in their treatments. At the same time, apart from the hospital wastewater samples, five samples were also collected from sewage drains in New Delhi in order to draw a comparison between hospital and community sources of antibiotic resistance. Only those drains that did not have hospitals in their catchment area were sampled. The tests and sampling were conducted during two different seasons in order to account for the seasonal variation in drug use and temperature differences. Sterilized containers with a capacity of 500 millilitres were used to collect these samples. They were transported to the laboratory in coolers, on ice and stored in refrigerators at a temperature of 4°C for analysis. Within 12 hours of collection, the samples were subjected to microbial culturing. Parameters such as pH, conductivity, temperature, total dissolved solids (TDS), and so on were quantified and measured using hand-held probes.

The following pathogenic parameters were quantified and studied based on the time scale (seasonal variation) and other factors – faecal coliforms (FC), total coliforms (TC), ESBL-resistant and carbapenem-resistant Enterobacteriaceae (CRE) abundance; selected ESBL-(blaCTX, blaOXA, blaTEM) and CRE-related (blaNDM-1) ARGs; and intl genes (as a reflection of HGT (horizontal gene transfer)). The data was analysed through bivariate correlation analysis and strong positive correlations were found between FC levels, and CRE ($r=0.924$; $p=0.005$), ESBL-resistant bacteria ($r=0.913, p=0.010$), TC ($r=0.921, p=0.008$), and blaNDM-1 gene ($r=0.934, p=0.009$) concentrations in hospital-originating sewage, from the entire city. Overall, the analysis revealed a close relationship between the abundance of faecal bacteria in hospital wastewaters, and the CRE and ESBL-resistance bacteria and genes levels that enter the surrounding ecosystem. The above-mentioned parameters were also observed to be higher in the hospital wastewaters during the winter season.

9.3.2 Central India

Diwan *et al.* (2012) conducted a study across two privately funded tertiary care hospitals situated in the district of Ujjain in Madhya Pradesh, located in central India. Around 276 samples were collected from wastewater generated at these hospitals in the period from 2008 to 2009. In order to confirm resistance to the antibiotics, the agar dilution method was used to determine the minimum inhibitory concentration (MIC) of cefotaxime and ciprofloxacin according to the Clinical & Laboratory Standards Institute (CLSI) breakpoints. Cefotaxime and ciprofloxacin are the most widely used expanded-spectrum cephalosporin and fluoroquinolone antibiotics at the hospital under study. The *E. coli* reference strain ATCC 25922 was used to perform quality control and the alkaline lysis method was used to extract total bacterial DNA from the isolated samples.

It was observed that out of the 32 *E. coli* isolates, a total of 27 of them showed resistance to cefotaxime. Out of these 27 samples, the presence of blaCTX-M-15 only or blaSHV-12 only confirmed that 25 of the isolates were ESBL-positive. Out of the 30 fluoroquinolone-resistant *E. coli* isolates that were under study, 90% of them carried the aac(6′)-Ib-cr ($n=27$) gene, 20% carried some other PMQR determinant genes (such as qnrA ($n=1$), qnrB ($n=2$) and qepA ($n=3$)) and none of the isolates carried the qnrS gene. Of the isolates, 21 showed the presence of both blaCTX-M-15 and aac(6′)-Ib-cr. Through phylogenetic grouping, it was discovered that 21 *E. coli* isolates belonged to the phylogenetic group B2, which denotes virulent extraintestinal strains, 7 isolates belonged to B1 and the remaining 2 belonged to group A (that denotes commensal isolates) and group D.

In conclusion, in spite of the fact that the number of samples collected was relatively low, the alarmingly high detection levels of *E. coli* that is resistant to expanded-spectrum cephalosporin and fluoroquinolone antibiotics in the wastewater samples collected from hospital wastewaters is of great concern and requires quick action (Alam *et al.* 2020).

9.3.3 Lucknow, Uttar Pradesh

Alam *et al.* (2020) conducted a study at the King George Medical University, Lucknow, in Uttar Pradesh in India. In order to carry out the study, samples were collected from three different wards at the King George Medical University in Lucknow between December 2015 and May 2016, for a duration of six months. Metal amended plates were used for isolation and tolerance of coliforms in the samples collected. The disc diffusion method was used to perform the antibiotic sensitivity test. Using the alkaline lysis method, the plasmid DNA of the bacterial isolates was further studied. The experiment's conjugation study was carried out both in a nutrient medium as well as wastewater.

For a majority of the isolates, their minimum inhibitory concentrations (MICs) were found to be at levels of roughly 400 to 1600 μg/ml against all the metals. The isolates showed more resistance against the following – methicillin, penicillin and cephradine. The least resistance was noticed to be towards gentamicin in *E. coli* and *Enterobacter.* The following six residues of amino acid – Glu104, Ala237, Tyr105, Gly238, Asn132 and Asn170 – of the beta-lactamase were observed. Plasmid DNA size range was observed to be 48–58.8 kb. Through the conjugation experiments, it was discovered that the conjugate samples showed higher transfer frequency (5.50-1 and 3.60-1) rates among the antibiotics and metals that were tested.

The findings of the study make it clear that hospital wastewaters pose a major challenge to public health. The predominant coliform species found in the study are now being linked to infectious outbreaks. This can be prevented by treating the wastewater before it is discharged from hospitals into the community sewage systems and public wastewater treatment units. Moreover, the wastewater collection systems and infrastructure also need to be improved in order to prevent leaks and direct contact of pathogenic wastewater with humans and animals.

9.4 LEGISLATION

According to the Indian government's mandate, the responsibility for managing and handling healthcare waste primarily falls on the healthcare facilities (HCFs) where the waste is generated. This includes activities that are performed by them in the community too. The generators of waste, that is, the HCFs, are also required to bear other responsibilities such as the segregation of waste materials, their collection from the rooms, the in-house transportation of wastes generated, performing pre-treatment procedures on the waste and, finally, storing the pre-processed waste. This waste is then collected by the operators of the common bio-medical waste treatment facility (CBWTF). Therefore, in order to ensure that the HCFs are managing their waste efficiently and properly, each category of the staff needs to not only understand, but also implement the waste handling procedure as laid out in the Biomedical Waste Management (BMWM) Rules, 2016.

Wastes produced by HCFs can be classified into the following three categories:

(1) Biomedical waste
(2) General waste
(3) Other wastes

Bio-medical waste has been defined as follows, 'any type of waste, which is generated during the diagnosis, treatment or immunization of human beings or animals or research activities pertaining to or in the production or testing of biological samples or at community health camps'. Examples of waste materials that fall under this category include all types of wastes that are generated at HCFs and if not disposed of properly, have the potential to cause ill effects to humans and the environment. All these harmful waste materials that can damage public health and the environment can be categorized as infectious and these types of waste materials are required to be handled in accordance with the Biomedical Waste Management Rules, 2016. This category of waste is known to comprise anywhere between 10% and 15% of the total waste generated by a HCF. Some examples of the types of wastes that fall under this category include the following – waste materials that contact a patient's blood, infected parts, body secretions, biological liquids such as chemicals, medicines, medical supplies, lab discharge, sharp metal objects and glass instruments, plastics and so on. The Biomedical Waste Management Rules, 2016, categorize HCF-generated bio-medical waste into four broad categories on the basis of the route of segregation that is required for them and assign a corresponding colour code. Various types of biomedical wastes that are generated at hospitals are put under one of the following headings:

- Category Yellow
- Category White
- Category Blue
- Category Red

Category yellow waste consists of a wide range of waste materials such as human and animal anatomical wastes, soiled wastes, discarded and expired medicines, chemical waste materials, mattresses, discarded linen, bedding that has been soiled by fluids such as blood or other bodily fluid, patients clothes such as gowns, masks, microbiological, biotechnological and other clinical laboratory waste, and waste that is pre-treated. The liquid waste that is generated during the use of chemicals in production of biological and used or discarded disinfectants, infected secretions, silver X-ray film developing liquid, discarded chemicals such as formalin, liquid from laboratories and floor cleaning and washing, house-keeping activities and disinfecting activities, body fluids that are collected during medical procedures, and so on also fall under this category. This also includes the hospital wastewater that is generated.

The effluents produced from HCFs are mandated to meet certain standards before they can be discharged into the common sewer:

(1) pH should range from 6.5 to 9
(2) upper limit on suspended solids is 100 mg/l

(3) upper limit on oil and grease is 10 mg/l
(4) BOD (biochemical oxygen demand) and COD (chemical oxygen demand) are required to be 30 mg/l and 250 mg/l, respectively.

The above stated limits are applicable to all hospitals and HCFs that have their sewage system linked to the public sewage system for the city, but the sewage does not get carried to the local wastewater treatment unit; or the hospital sewage discharge system is not connected to the public sewage system at all.

If the hospital's sewage network is linked to the city's sewage network and that leads to the municipal wastewater treatment unit, then the general standards of sewage discharge (as specified under the Environment (Protection) Act, 1986) shall be applied.

For complying with the standards laid out in the Biomedical Waste Management Rules, 2016, every hospital is required to have an effluent treatment plant (ETP) to treat the wastewater that is generated in these facilities. Typical sources from which wastewater is generated at hospitals include – laboratories, wards, floor washing, used disinfectants, hand washing, washing of patients' area, laundry, accidental spillage disposal, firefighting, bathrooms/toilets and so on. The liquid waste that is generated during activities such as use of chemicals or discarded disinfectants, cleaning, and house-keeping, as well as liquid from laboratories, floor washings, disinfecting activities, aspirated body fluids, and infected secretions is required to be separately collected and provided with some level of pre-treatment, before it is mixed with the other wastewater generated from the facility.

The total collected wastewater should then be sent to the effluent treatment plant, where the treatment process can be categorized as follows:

(1) Primary treatment – this includes preliminary treatment processes such as equalization, neutralization, precipitation and clarification.
(2) Secondary treatment – this includes a high-rate-aerobic treatment process followed by secondary settling in a settling tank.
(3) Tertiary treatment – this comprises the final treatment processes including pressure filtration, disinfection and disposal to the common drain/sewer.

If efficiently treated, the treated water can also be reused in some systems. Some of the major options for reuse of treated wastewater include flushing, horticulture and scrubbing. However, the wastewater generated from hospital premises is typically treated in the effluent treatment plant and then disposed of into the drain.

According to the Biomedical Waste Management Rules of 2016, the hospital's chemical liquid waste is required to be collected using a different system for collection, to enable provision of some pre-treatment. Hospitals with large stand-alone laboratories are required to have a separate drainage system that leads to the set-up provided for pre-treatment before liquid waste is mixed with the rest of the wastewater from other parts of the hospital for future treatment. For small and mid-scale HCFs that do not have a different system of drainage or collection, the liquid waste is first collected on-site, using containers, for pre-treatment and it is then mixed with other wastewaters. The fluid used in silver

X-ray film developing has to be either given or sold to the authorized recyclers for recovery of resources. If that is not possible, this waste has to be handed over to the Common Bio-medical Waste Treatment and Disposal Facility (CBWTF) as yellow (e) category chemical waste. Although pre-treatment procedures vary from one chemical effluent to another, typical pre-treatment includes processes such as neutralization or precipitation, followed by disinfection of the chemical wastewater prior to being mixed with the rest of the wastewater from other parts of the hospital facility. Before being mixed with the common hospital wastewater, disinfection is a key process. Preferably, disinfection of the water has to be carried out by exposing the effluent to ultraviolet light to sterilize it. Another alternative is disinfection through use of chemicals. However, this is discouraged because these chemicals can create microbiological problems later on in the treatment process, thereby hindering efficient and cost-effective treatment (Central Pollution Control Board, Ministry of Environment, Forest & Climate Change, 2016).

9.5 RECOMMENDATIONS

It is a well-known fact that improper and unscientific disposal of hospital wastewater not only pollutes the environment, but is also a significant threat to public safety. Pharmaceuticals also pose a serious threat to marine and aquatic ecosystems. Moreover, conventional wastewater treatment plants in several municipalities across the world are not equipped to deal with contaminants that are discharged from hospitals. This reduces the treatment efficiency and can also damage the existing treatment infrastructure. At the same time, clean potable water is an absolute necessity that is required by hospitals to safeguard hygiene and health of their patients. The hospital wastewater treatment infrastructure is lacking in the majority of hospitals in India. As a developing country, the government's expenditure on health care infrastructure is inadequate and therefore wastewater is not being treated effectively. In spite of government mandated guidelines and standards, the treatment procedures (if any) vary from one region to another, as we have already observed in the case studies. In order to improve the overall handling of hospital wastewater in India, some of the key recommendations include the following:

(1) Improved on-site treatment – given that the wastewater generated at hospitals has different characteristics from the wastewater flowing through community sewers, it is important that the water is given some form of pre-treatment before it is discharged into the community sewers (Kozub *et al.*, 2021). The guidelines stated in the Biomedical Waste Management Rules of 2016 also state that the chemical liquid waste should be pre-treated before being sent to the effluent treatment unit, from where the water can be discharged into sewers. In order to ensure this, hospitals and HCFs need to be upgraded with installation of pre-treatment units in-house, where the waste water can be treated. Some of the common processes that are used in pre-treatment of wastewater include neutralization of water through special tanks and precipitation of the wastewater constituents and their removal from the effluent. In addition to this, thorough disinfection of the wastewater is also important for ensuring that the pathogenic threats can be reduced.

UV-based disinfection is better than chemical disinfection. This is because the addition of chemicals can alter the chemical properties of wastewater which in turn can create issues in the downstream treatment processes. Performing on-site treatment can greatly reduce the threat of the general public's exposure to wastewater, antibiotics and pathogens. This also protects the wastewater treatment infrastructure downstream.

(2) Improved wastewater collection infrastructure – one of the biggest threats of improper hospital wastewater disposal is the exposure of pathogens to pharmaceuticals and antibiotics Vambol *et al.*, 2021c. This in turn leads to antibiotic resistance and that can be a serious public health concern, as has been discussed earlier. Apart from this, the exposure of humans to pathogens discharged from hospitals is also a serious issue that can lead to spread of diseases. Marginalized and poor communities who reside in slums are especially at risk of such pathogenic breakouts, because of their exposure to open drains. In order to prevent this, the infrastructure needs a major revamp. The transmission of hospital wastewaters to common treatment units needs to be made secure and leak proof. The open drains should be closed and the cleaning processes should be automated. Practices such as manual scavenging, which is rampant in India, need to be discontinued and replaced with better technological advancements. In addition to this, the general public also needs to be made more aware regarding disposal of expired medicines, tablets and antibiotics. Although on-site treatment solutions are the most effective, the improvement of existing wastewater infrastructure can be carried out at lesser costs by the government.

(3) Stringent enforcement of guidelines and standards – in spite of government mandated rules laid down in the Biomedical Waste Management Rules of 2016, the majority of Indian HCFs are not meeting the standards that have been stipulated. Several hospitals run without dedicated effluent treatment plants and pre-treatment units and this is a leading cause of the negative effects of hospital wastewater on public health and the environment. In order to prevent this, the rules should be enforced more strictly. Hospitals that do not meet standards should be penalized and fined. Regular audits and checks should be conducted by local municipal bodies. The hospitals funded by the state and the central governments should prioritize the creation and utilization of sound wastewater handling infrastructure (Carraro *et al.*, 2016).

REFERENCES

Al Aukidy M., Verlicchi P. and Voulvoulis N. (2014). A framework for the assessment of the environmental risk posed by pharmaceuticals originating from hospital effluents. *Science of the Total Environment*, **493**, 54–64, https://doi.org/10.1016/j.scitotenv.2014.05.128

Alam M., Imran M. and Ahmad S. S. (2020). Screening of metal and antibiotic resistance in Beta-lactamase producing coliform Bacteria from hospital wastewater of northern India. *Recent Patents on Biotechnology*, **14**(1), 63–77, https://doi.org/10.2174/1872208313666191002130406

Balwani K. S. and Nagarnaik P. B. (2017). Water and waste water management of a hospital-a review. *International Journal of Science and Research*, **6**(4), 1414–1417.

Bureau of Indian Standards. (2010). Code of basic requirements for water supply, drainage and sanitation.

Carraro E., Bonetta S., Bertino C., Lorenzi E., Bonetta S. and Gilli G. (2016). Hospital effluents management: chemical, physical, microbiological risks and legislation in different countries. *Journal of Environmental Management*, **168**, 185–199, https://doi.org/10.1016/j.jenvman.2015.11.021

Central Pollution Control Board, Ministry of Environment, Forest & Climate Change. (2016). Guidelines for Management of Healthcare Waste as per Biomedical Waste Management Rules. https://mpcb.gov.in/sites/default/files/biomedical-waste/ManagementofHealthcareWaste03032020.pdf (last accessed on August 4, 2021)

Central Public Health and Environmental Engineering Organisation (CPHEEO). (1993). Manual on sewerage and sewerage treatment.

Chandra H. (1999). Hospital waste: an environmental hazard and its management. *C. Newsletter of ISEB, India*, **5**(3), 80–85.

Desbiolles F., Malleret L., Tiliacos C., Wong-Wah-Chung P. and Laffont-Schwob I. (2018). Occurrence and ecotoxicological assessment of pharmaceuticals: is there a risk for the Mediterranean aquatic environment?. *Science of the Total Environment*, **639**, 1334–1348, https://doi.org/10.1016/j.scitotenv.2018.04.351

Diwan V., Chandran S. P., Tamhankar A. J., Stålsby Lundborg C. and Macaden R. (2012). Identification of extended-spectrum β-lactamase and quinolone resistance genes in Escherichia coli isolated from hospital wastewater from central India. *Journal of Antimicrobial Chemotherapy*, **67**(4), 857–859, https://doi.org/10.1093/jac/dkr564

Environmental Protection Agency, Watersense. (2012). Saving water in hospitals. https://www.epa.gov/sites/default/files/2017-01/documents/ws-commercial-factsheet-hospitals.pdf (last accessed on August 4, 2021)

Gupta P., Mathur N., Bhatnagar P., Nagar P. and Srivastava S. (2009). Genotoxicity evaluation of hospital wastewaters. *Ecotoxicology and Environmental Safety*, **72**(7), 1925–1932, https://doi.org/10.1016/j.ecoenv.2009.05.012

Hoek A. C., Malekpour S., Raven R., Court E., and Byrne E. (2021). Towards environmentally sustainable food systems: decision-making factors in sustainable food production and consumption. *Sustainable Production and Consumption*, **26**, 610–626.

Huang H., Zeng S., Dong X., Li D., Zhang Y., He M. and Du P. (2019). Diverse and abundant antibiotics and antibiotic resistance genes in an urban water system. *Journal of Environmental Management*, **231**, 494–503, https://doi.org/10.1016/j.jenvman.2018.10.051

Jolibois B. and Guerbet M. (2006). Hospital wastewater genotoxicity. *Annals of Occupational Hygiene*, **50**(2), 189–196.

Kataki S., Chatterjee S., Vairale M. G., Sharma S. and Dwivedi S. K. (2021). Concerns and strategies for wastewater treatment during COVID-19 pandemic to stop plausible transmission. *Resources, Conservation and Recycling*, **164**, 105156, https://doi.org/10.1016/j.resconrec.2020.105156

Khan A., Aziz H., Khan N. A., Hasan M., Ahmed S., Farooqi I., Dhingra A., Vambol V., Changani khorasgani F., Yousefi M., Islam S., Mozaffari N. and Mahtab M. (2021). Impact, disease outbreak and the eco-hazards associated with pharmaceutical residues: a Critical review. *International Journal of Environmental Science and Technology*. doi: 10.1007/s13762-021-03158-9.

Kozub P., Kozub S., Mozaffari N., Sihag P., Mozaffari N., Vambol S., ... and Khan N. A. (2021). Treatment schemes–conventional and dedicated for PhACs treatment. *Pharmaceutical Wastewater Treatment Technologies*, **181**, 181–220. https://doi.org/10.2166/9781789061338_0181

Kumari A., Maurya N. S. and Tiwari B. (2020). Hospital wastewater treatment scenario around the globe. *Current Developments in Biotechnology and Bioengineering*. 549–570. doi: 10.1016/B978-0-12-819722-6.00015-8. Epub 2020 May 1. PMCID: PMC7252247.

Lamba M., Graham D. W. and Ahammad S. Z. (2017). Hospital wastewater releases of carbapenem-resistance pathogens and genes in urban India. *Environmental Science & Technology*, **51**(23), 13906–13912, https://doi.org/10.1021/acs.est.7b03380

Lee S., Kim Y., Park J., Shon H. K. and Hong S. (2018). Treatment of medical radioactive liquid waste using forward osmosis (FO) membrane process. *Journal of Membrane Science*, **556**, 238–247, https://doi.org/10.1016/j.memsci.2018.04.008

Mubedi J. I., Devarajan N., Le Faucheur S., Mputu J. K., Atibu E. K., Sivalingam P., … and Poté J. (2013). Effects of untreated hospital effluents on the accumulation of toxic metals in sediments of receiving system under tropical conditions: case of south India and democratic Republic of Congo. *Chemosphere*, **93**(6), 1070–1076, https://doi.org/10.1016/j.chemosphere.2013.05.080

Ooi G. T., Tang K., Chhetri R. K., Kaarsholm K. M., Sundmark K., Kragelund C., … and Andersen H. R. (2018). Biological removal of pharmaceuticals from hospital wastewater in a pilot-scale staged moving bed biofilm reactor (MBBR) utilising nitrifying and denitrifying processes. *Bioresource Technology*, **267**, 677–687, https://doi.org/10.1016/j.biortech.2018.07.077

Sharma P., Mathur N., Singh A., Sogani M., Bhatnagar P., Atri R. and Pareek S. (2015). Monitoring hospital wastewaters for their probable genotoxicity and mutagenicity. *Environmental Monitoring and Assessment*, **187**(1), 1–9.

Vambol S., Vambol V., Khan N., Mozaffari N. and Mozaffari N. (2021a). Algorithm for Elimination of the Consequences of the Oil-Polluted Water. Scientific works of the Third International Scientific and Practical Conference Modern Trends in the Development of Information Systems and Telecommunication Technologies, January 25–26, 2021, Kyiv, Ukraine. NUHT.

Vambol S., Vambol V., Mozaffari N., Mahmoud A. E. D., Ramsawak N., Mozaffari N., … and Khan N. A. (2021b). Comprehensive insights into sources of pharmaceutical wastewater in the biotic systems. *Pharmaceutical Wastewater Treatment Technologies*, **17**, 17–46, https://doi.org/10.2166/9781789061338_0017

Vambol V., Khan N. A., Vambol S. and Ahmed S. (2021c). Future scope–expected treatments for phcs in future. *Pharmaceutical Wastewater Treatment Technologies*, **375**, 375–384. https://doi.org/10.2166/9781789061338_0375

Van Boeckel T. P., Gandra S., Ashok A., Caudron Q., Grenfell B. T., Levin S. A. and Laxminarayan R. (2014). Global antibiotic consumption 2000 to 2010: an analysis of national pharmaceutical sales data. *The Lancet Infectious Diseases*, **14**(8), 742–750, https://doi.org/10.1016/S1473-3099(14)70780-7

Van Hoek A. H., Mevius D., Guerra B., Mullany P., Roberts A. P. and Aarts H. J. (2011). Acquired antibiotic resistance genes: an overview. *Frontiers in Microbiology*, **2**, 203, https://doi.org/10.3389/fmicb.2011.00203

Wang Q., Wang P. and Yang Q. (2018). Occurrence and diversity of antibiotic resistance in untreated hospital wastewater. *Science of the Total Environment*, **621**, 990–999, https://doi.org/10.1016/j.scitotenv.2017.10.128

Yang Y., Shi W., Lu S.-Y, Liu J., Liang H., Yang Y., Duan G., Li Y., Wang H. and Zhang A. (2018). Prevalence of antibiotic resistance genes in bacteriophage DNA fraction from Funan River water in Sichuan, China. *Science of The Total Environment*, **626**, 835–841, https://doi.org/10.1016/j.scitotenv.2018.01.148.

doi: 10.2166/9781789062625_0183

Chapter 10

Hospital wastewater treatment in Nepal: Status, challenges and future perspectives (case studies of effluent management in Nepal)

Amrita Gautam[1,2], Krishna Ram Yendyo[3], Prabina Shrestha[4], Isha Manandhar[3], Sagar Kunwar[2,5], Sudan Panthi[6] and Anish Ghimire[7]*

[1]*Faculty of Spatial Development and Infrastructure Systems, Technical University of Cologne, Germany*
[2]*drinkPani, Kathmandu, Nepal*
[3]*Eco Concern Pvt. Ltd, Lalitpur, Nepal*
[4]*Environment and Public Health Organization, Kathmandu, Nepal*
[5]*La Grandee International College, Pokhara University, Nepal*
[6]*World Health Organization, Nepal*
[7]*Resource Recovery Research Group (Re3G), Department of Environment Science and Engineering, Kathmandu University, Nepal*
Correspondence: amritagtm4@gmail.com

ABSTRACT

The population growth has significantly increased the number of health facilities in Nepal, which has contributed to an increase in the generation of hospital waste and wastewater in Nepal. There are limited studies in the management of this generated waste and wastewater. This research aims to present the status of the characteristics of hospital wastewater, its treatment status, concerns and treatment challenges, and future perspectives for Nepal through the detailed review of past literature and research conducted by the organizations working in this sector. The total number of healthcare facilities (including both public and private healthcare facilities) in Nepal has reached 6934 with an estimated waste generation in the range of 0.99 to 1.73 kg/bed/day. But there is no data for wastewater generated from the health facilities. Most of the hospital wastewater is discharged into public sewers and only a few hospitals have effluent treatment plants. The hospitals have been using mostly decentralized wastewater treatment systems, which can reduce some conventional parameters in the effluent to the limit defined by the national standards. Since many hospitals are disposing of their waste and effluent without treatment into water bodies, there is a public health risk from pathogens, pharmaceutically active compounds, and other emerging pollutants. The major challenges in the adoption of appropriate treatment technologies and their management in hospitals to increase treatment include strong enforcement of the national effluent discharge standards, lack of

Editor(s): Nadeem A Khan, Viola Vambol, Sergij Vambol, Nastaran Mozaffari and Niloofar Mozaffari

adequate budget allocation, location, proper organizational management, data availability and effective monitoring. With proper governmental monitoring and support, there are opportunities to effectively manage waste/wastewater in hospitals in Nepal.

Keywords: Hospital wastewater (HWW), public health, techniques, treatment, wastewater management

CONTENTS

10.1 INTRODUCTION

Rapid urban growth, together with the increasing population, has caused the surge in wastewater generation in Nepal. Everyday industries, agriculture and the general population use water as per their demand and release different types of hazardous compounds in their wastewater. Since the evolution of wastewater treatment, there has been lots of research done for the elimination of macropollutants from hospital wastewater (HWW), industrial wastewater (IWW) and urban wastewater (UWW) (Verlicchi *et al.*, 2010). Depending on the existing regulations in different countries, HWW is often assimilated to UWW, and discharged into municipal sewerage and directed to a wastewater treatment plant (WWTP) where for co-treatment with urban and industrial effluents. Here the assumption is that all the wastewaters are similar in terms of pollutants, concentration, and loads and dilution of HWW and IWW with a large volume of UW (Verlicchi *et al.*, 2010). It is likely that these assumptions are not correct because these wastewaters are different. Referring to the conventional parameters, and by considering their usual concentrations in the influents to municipal WWTPs, it is observed that in HWWs biological oxygen demand (BOD_5), chemical oxygen demand (COD) and suspended solids (SS) are 2–3 times higher than in UWW (Verlicchi *et al.*, 2010). Moreover, HWW is characterized by the presence of different micropollutants, such as pharmaceutically active compounds (PhACs), several pathogens including antibiotic residues, antibiotic-resistant microorganisms and persistent viruses, among others (Song *et al.*, 2020). Hospital wastewater in Nepal is being discharged directly or indirectly into public sewers, which releases into the river and its tributaries, mostly without any treatment. In the past, there have been few centralized treatment plants in Kathmandu to treat all the wastewater from the core area of the cities. However, they continued to not function well due to high capital investment required, high operation and maintenance costs, scarcity of spare parts and lack of skilled workforce (Jha & Bajracharya, 1998). Five centralized treatment plants were installed in Kathmandu between 1977 and 1999 with a total capacity of 36.3 MLD (million litres per day) (Rajendra *et al.*, 2016). Most of these treatment plants were lagoon-based treatment technology. These treatment plants failed and became dysfunctional many years ago. Due to the centralized treatment plants being non-operational, the major river tributaries in Kathmandu valley remain polluted as all the hospitals, industries, office buildings and residents discharge untreated sewage through the sewer network into the rivers. Nevertheless, in recent years, Nepal government has begun upgrading old and building new treatment plants, at Guheshwori, Kodku, Sallaghari, Dhobighat and Gokharna, with advance treatment technologies with a combined capacity of 90.5 MLD (Asian Development Bank, 2013).

The centralized wastewater treatment plants can only serve the areas which have their catchment linked to the sewer networks. At present, the metropolitan and sub-metropolitan cities, and a few municipalities in Nepal have access to sewer networks in the majority of their areas. Among 18 municipalities in Kathmandu valley, only 5 are well connected with the sewer networks while the rest are relying on on-site treatment systems (CWISTA Hub, 2019). There are hundreds of healthcare facilities, including polyclinics, health posts, and small and large hospitals, which are not linked to sewer networks. Some of the communities and hospitals that lack connection to sewer networks are treating the wastewater from their buildings with decentralized wastewater treatment systems (DEWATSs) based on wetland technology (Green *et al.*, 2003). By 2014, there were more than a dozen wetland-based wastewater treatment technology systems installed in Nepal and observed to be properly functioning. The DEWATS is now being considered as an effective, economic and environmentally friendly wastewater treatment system in Nepal and has been adopted by many of Nepal's community hospitals (Jha & Bajracharya, 1998).

With the increasing population and number of healthcare facilities (HCFs), healthcare waste (HCW) is posing growing problems due to the limitations of technological and financial tools to ensure its proper management (Pathak *et al.*, 2021). There are so many potential hazards associated with handling HCW, posing risks not only to the patients and healthcare workers but also to the community and the environment (World Health Organization [WHO], 2005).

10.2 TOOLS AND METHODS

The study was conducted based on a literature review of previous studies, case reports and unpublished reports on hospital waste and effluent treatment in Nepal. The secondary data from the research carried out in 16 hospitals by EcoConcern Nepal on hospital healthcare wastes management (HCWM), and the monitoring studies conducted by the Environment and Public Health Organization (ENPHO) on two hospital wastewater treatment plants (Dhulikhel Hospital DEWATS and Sushma Koirala Memorial Hospital DEWATS) were also considered in the chapter. Moreover, the design reports of Dhulikhel Hospital DEWATS prepared by ENPHO were also reviewed. The monitoring information on characteristics of HWW was compared with the National Standard on Hospital Effluent.

10.3 HOSPITAL WASTE AND WASTEWATER MANAGEMENT SCENARIO IN NEPAL

There has been a substantial increase in the number and capacity of HCFs; the total number of HCFs (including both public and private HCFs) in Nepal has reached 6934, and one-third of these facilities are in Bagmati Province (DoHS, 2019). Most of the larger hospitals are in Kathmandu Valley and few are outside the valley. There are 366 private hospitals in Nepal, among which 99 are in Kathmandu Valley, with around 123 government hospitals all over the country running under the Ministry of Health and Population (MOHP). The ever-increasing number of HCFs has led to complications in the management of HCW and HWW (Joshi *et al.*, 2017).

10.3.1 Healthcare waste management scenario

Recent published studies revealed that the total HCW from Nepal is estimated at 0.99 to 1.73 kg/bed per day (HCWM SOP, 2020). Environmental health, and HCWM are under the jurisdiction of the Department of Health Services (DoHS) within the Ministry of Health. DoHS is responsible for monitoring the quality of drinking water, waste materials and air quality, and determining criteria for quality in the work associated with the Union, State, and local government. Though the section is responsible for HCWM, the responsibility of HWW management has not been mentioned clearly in the role.

Nepal's HCWM guidelines (MOHP, 2014) state that every HCF should have an HCWM committee to oversee its waste but a study conducted by Eco Concern Pvt. Ltd. (also EcoConcern) in 16 private hospitals of Kathmandu indicated that only 56% of hospitals had HCWM committees (EcoConcern, 2019). Many hospitals did not have clear rules and protocols, guidelines, and organizational structures to deal with HCW. The study revealed that lack of infrastructure for proper HCWM, manpower, financial resources, HCWM knowledge and guidelines, and unavailability of required items (e.g., colour coded buckets) were mentioned as the major reasons for not having an HCWM committee. In many cases, HCWM is often left to hospital sweepers and cleaners (EcoConcern, 2019).

Out of 16 surveyed hospitals, only 16% have written policies, plans, manuals, or procedures dealing with HCMW. Whereas the remaining 84% of hospitals do not have any written plans and policies of their own, so they have followed and enforced the national HCWM guidelines as per their convenience and feasibility. Effective HCWM practices depend on the priority assigned to them by the hospital management.

The study also revealed that in most of the hospitals there is no trained or qualified human resource available for HCWM due to which the hospital sweepers are given the responsibility of healthcare waste handling as well as management which shows that there is still a lack of proper waste management in the facility from policy level to operational level (EcoConcern, 2019).

Since hospitals are sensitive areas, infection prevention is very much in need and waste management plays a vital role in preventing such infections. As per the survey, the main problems faced during the application of National HCWM regulation were lack of resources and proper manuals to follow guidelines, difficulty in finding proper colour-coded buckets for collection of wastes and so on.

Most hospitals do not have a proper system of tracking and recording the generated waste. Though the wastes are segregated at the source, it was found that in a few of the hospitals the wastes of high risk were found to be mixed with those of low to no risk (EcoConcern, 2019). This gave rise to concern of possible human health and environmental risks. For those hospitals that did treat their waste, the common mode of treatment was either using autoclaves and/or incinerators but, from the survey, it was found that even though the HCF's had autoclaves, due to lack of resources personal treatment was not carried out.

10.3.2 Hospital wastewater management scenario

The Department of Environment, Ministry of Forest and Environment is responsible for monitoring overall environmental issues including HWW management in Nepal. Nepal has implemented effluent standards for several industries, and in 2003 the Government of Nepal, Ministry of Population and Environment promulgated ad hoc Generic Standards Part III Tolerance Limits for wastewater to be discharged into inland surface waters from the combined wastewater treatment plant (Nepal Gazette, 9/3/2060 Nepali Calendar). Similarly, on 5th August 2019 (Nepal Gazette, 18/4/2060 Nepali Calendar), hospital wastewater effluent standards were published. This set the limits for parameters such as BOD_5 – 100 mg/l, COD – 250 mg/l, mercury – 0.01 mg/l, lead – 0.1 mg/l, and faecal coliform – 400 MPN/10 ml among others in the HWW effluent to be discharged. More details on the effluent limits are presented in Table 10.1.

The National Water Supply and Sanitation Policy 2014 (draft) indicates that all hospitals and corporate buildings need their effluent to be treated to meet required standards before discharging it into public sewers (Ministry of Urban Development [MoUD], 2014). However, only a few hospitals are equipped with their own effluent treatment plant (ETP) to treat the wastewater before discharging it into the sewer networks or inland. It was found that almost 89% of

Table 10.1 Hospital effluent standard in Nepal.

Parameters	Concentration not to Exceed
pH	5.5–9
Suspended solids, mg/l	100
BOD_5, mg/l	100
COD, mg/l	250
Oil and grease, mg/l	10
Phenolic compound, mg/l	1
Total residual chlorine, mg/l	1
Mercury (Hg), mg/l	0.01
Cyanide (CNCn), mg/l	2
Lead (Pb), mg/l	0.1
Faecal coliform, MPN/100 ml	400

the hospitals opted for no prior actions of treatment of the effluent before direct disposal into the municipal sewerage system or inland water bodies, though the effluent discharge standards exist. Most of the hospitals do not treat their effluent before discharging it into public sewers or other water bodies or inland. Though most surveyed hospitals have some system for managing solid waste, only a few hospitals (19%) treat their liquid waste before disposal. A very small fraction of these hospitals have DEWATS for HWW treatment (EcoConcern, 2019).

Different wastewater treatment technologies are available to efficiently treat conventional parameters and micropollutants such as PhACs, surfactants, mercury and so on. present in wastewater. The physico-chemical treatments, biological methods (based on an activated sludge processor or ASP), membrane processes such as nano-filtration, reverse osmosis, ozonation and other advanced oxidation processes, and disinfections are frequently applied in developed countries for the treatment of HWW. Among the treatment technologies, constructed wetlands (CWs) combined with physical treatment (primary) methods are commonly used as decentralized wastewater treatment options in many communities and institutions such as hospitals in Nepal. However, the conventional wastewater treatment plants are not designed to remove the emerging pollutants of concern which are often present in HWW.

During the ongoing pandemic of SARS-CoV-2, there are obvious concerns about the presence of viruses in the HWW. As such, regarding the management of wastewater from HCFs, during the COVID-19 emergency, interim guidance prepared by MoHP (2020) mentioned that hospital wastewater should be treated in well-designed and well-managed wastewater treatment plants. The recent WHO interim guidance (WHO, 2020), explains that a surrogate human coronavirus survived for only 2 days in dechlorinated tap water and in hospital wastewater at 20°C. A waste stabilization pond (an oxidation pond or lagoon) is generally considered a practical and simple wastewater treatment technology particularly well suited to destroying pathogens, as relatively long retention times (20 days or longer) combined with sunlight, elevated pH levels, biological activity, and other factors serve to accelerate pathogen destruction. A final disinfection step may be considered if existing wastewater treatment plants are not optimized to remove viruses. Nepal government has legislation, policy, and guidelines for healthcare waste management (HCWM) in place for treatment and disposal of infectious waste from HCFs. However, compliance with these guidelines was a challenge even before the COVID-19 pandemic, and it is now becoming more and more challenging to manage huge amounts of infectious wastes (MOHP, 2020).

Many efforts have been directed towards proper and safe management of hazardous healthcare waste in Nepal by different organizations, both public and private. However, inadequate management practices are often implemented in most healthcare facilities. In this regard, there are many opportunities for the private sector in HCWM as there is a demand to improve service delivery that cannot be met by the public sector alone.

10.4 CASE STUDIES OF HWW TREATMENT

10.4.1 Dhulikhel Hospital DEWATS

Dhulikhel Community Hospital, a Kathmandu University hospital located in Dhulikhel Municipality in Nepal was established in 1996. At present, the hospital serves 476 beds with a 46% occupancy rate and is targeting to reach 750 beds in the near future. The total estimated water consumption in the hospital is 200 m^3/d mainly used in providing health services and laundry. Most of the effluent generated from the hospital buildings is connected to the common collection system, except for the hostel, and forensic buildings which are connected to two individual septic tanks (ENPHO, 2017).

10.4.1.1 Hospital effluent characteristics

The hospital effluent consists of harmful contaminants such as pathogens along with PhACs, heavy metals contributed from HCFs, toilets, and laundry. The characterization of hospital effluent conducted by ENPHO in three different years in Dhulikhel Hospital is shown in Table 10.2. The effluent was sampled at the inlet of the wastewater treatment plant in 2010. The data from 2017 represents the composite sample of effluents from the main building, outpatient departments, and maternity ward in the hospital while that of 2021 was sampled at the inlet of the upgraded wastewater treatment plant.

The results show that the biological parameters such as BOD_5, COD and TSS are higher than the effluent standard in Nepal in all three years. The BOD_5 and COD in effluent increased in 2017 compared to 2010 due to increased hospital capacity. However, it is lower in 2021 than in 2017 which might be due to the mixing of effluent from all buildings in the combined sewer before it enters the treatment plant. The heavy metals were also tested in the effluent in 2017, which traced the presence of Fe, Mn, Cu, Zn and Hg with Fe and Mn contents higher than the standard values.

10.4.1.2 Effluent treatment plant

Dhulikhel Hospital has been treating its effluent in constructed wetlands based on DEWATS since 1997, a year after its establishment. The treatment plant at

Table 10.2 Characteristics of effluent generated in Dhulikhel Hospital in different years.

Parameters	Unit	2010*	2017**	2021***
BOD_5	mg/l	60	357	220
COD	mg/l	432	690	540
pH		7.1	7.8	8.23
Total suspended solids (TSS)	mg/l	55	ND	164
Total phosphorus (TP)	mg/l	2	ND	ND
Total nitrogen (TN)	mg/l	19.5	ND	ND
Oil and grease	mg/l	3.6	ND	ND
Ag	mg/l	ND	0	ND
Fe	mg/l	ND	0.22	ND
Mn	mg/l	ND	0.06	ND
Cu	mg/l	ND	0.02	ND
Zn	mg/l	ND	0.1	ND
Co	mg/l	ND	0	ND
Ni	mg/l	ND	0	ND
Cd	mg/l	ND	0	ND
Cr (VI)	mg/l	ND	0	ND
Pb	mg/l	ND	0	ND
Hg	mg/l	ND	0.001	ND
As	mg/l	ND	0	ND

ND, no data.
*data source (ENPHO, 2015).
**data source (ENPHO, 2017).
***data source (ENPHO, 2021).

Table 10.3 Components of effluent treatment plant in Dhulikhel Hospital over time (ENPHO, 2015).

1997	2008	2020 (ENPHO, 2020)
Design flow: 25 m³/day Treatment units • 1 settler • 1 horizontal reed bed • 1 vertical reed bed	Design flow: 90 m³/day Treatment units • 1 settler • 1 baffled septic tank • 2 horizontal reed beds • 2 vertical reed beds • 1 sludge drying bed	Design flow: 100 m³/day Treatment units • 1 screen chamber • 1 cyclone • 1 settlery • 1 integrated anaerobic baffled reactor and anaerobic filtery • 1 horizontal flow constructed wetland • 1 sludge drying bed • 1 grit disposal unit

the hospital has undergone several upgrades and changes in the last 20 years of its establishment as shown by the salient features of the treatment plant in Table 10.3. The treatment plant was first introduced as a 'Biological Wastewater Treatment Plant' with the technical support of ENPHO, IWGA-SIG and the Institute of Water Provision Water Ecology and Waste Management, University Boku, Austria to treat 25 m^3/day (ENPHO, 2015). The plant consisted of a reed bed system with one settler, one horizontal flow reed bed (HFRB) and one vertical flow reed bed (VFRB). The system was effectively operated until 2007. With the increasing effluent volume due to increased hospital capacity, it was further upgraded in 2007 to 90 m^3/day capacity with the technical support of ENPHO (ENPHO, 2015). One baffled septic tank, one HFRB, one VFRB and one sludge drying bed were added to the previous system to treat the increased volume of effluent. After serving for a decade, this treatment system was over its design period and an assessment was conducted to determine whether to either rehabilitate or construct a new treatment plant to meet the demands of increasing effluent volume, considering the planned future upgrade in the hospital capacity. A new DEWATS treatment plant was proposed in 2017. The treatment plant with a capacity of 200 m^3/d was proposed to be constructed in two phases, each phase having 100 m^3/d capacity (ENPHO, 2017). Three years later, in 2020, Phase 1 of the treatment plant came into operation to treat 100 m^3/d of the hospital effluent.

The layout of the proposed new treatment plant in full phase is shown in Figure 10.1. The treatment plant was proposed with one screen chamber to prevent floating materials from entering the system; one cyclone to remove grit from the effluent; two settlers to stabilize and partially treat the effluent; two integrated anaerobic baffled reactor and anaerobic filters to degrade the organic matter in anaerobic condition; one horizontal flow constructed wetland (HFCW) and one vertical flow constructed wetland (VFCW) to further reduce BOD, nitrates and odour from the effluent; and a polishing pond before discharging the effluent to the nearby open drain.

At present, the first phase of the treatment plant is in operation which consists of an inlet chamber, cyclone, one settler, one integrated anaerobic baffled reactor and anaerobic filter, one horizontal flow constructed wetland and one sludge drying bed as shown in Figure 10.2.

10.4.1.3 Performance efficiency

The treatment plant in the hospital has been working effectively to remove the conventional pollutants, such as BOD_5, COD and TSS from the hospital effluent. The study conducted in 2010 shows that the treatment plant had removal efficiency of 90%, 48%, and 91% for BOD, COD and TSS, respectively (Figure 10.3).

The treatment plant was not fully functional in its later years (in 2018) with the increased amount of effluent and management of the units and thus, the efficiency gradually decreased over the years. The study conducted by Dugar (2018) shows the significant drop in the removal of BOD and COD in 2018 (see Table 10.4) in comparison to data shown in Figure 10.3.

The new treatment plant at the hospital was constructed in 2019 and began operation in 2020. The monitoring conducted in 2021 in this new treatment

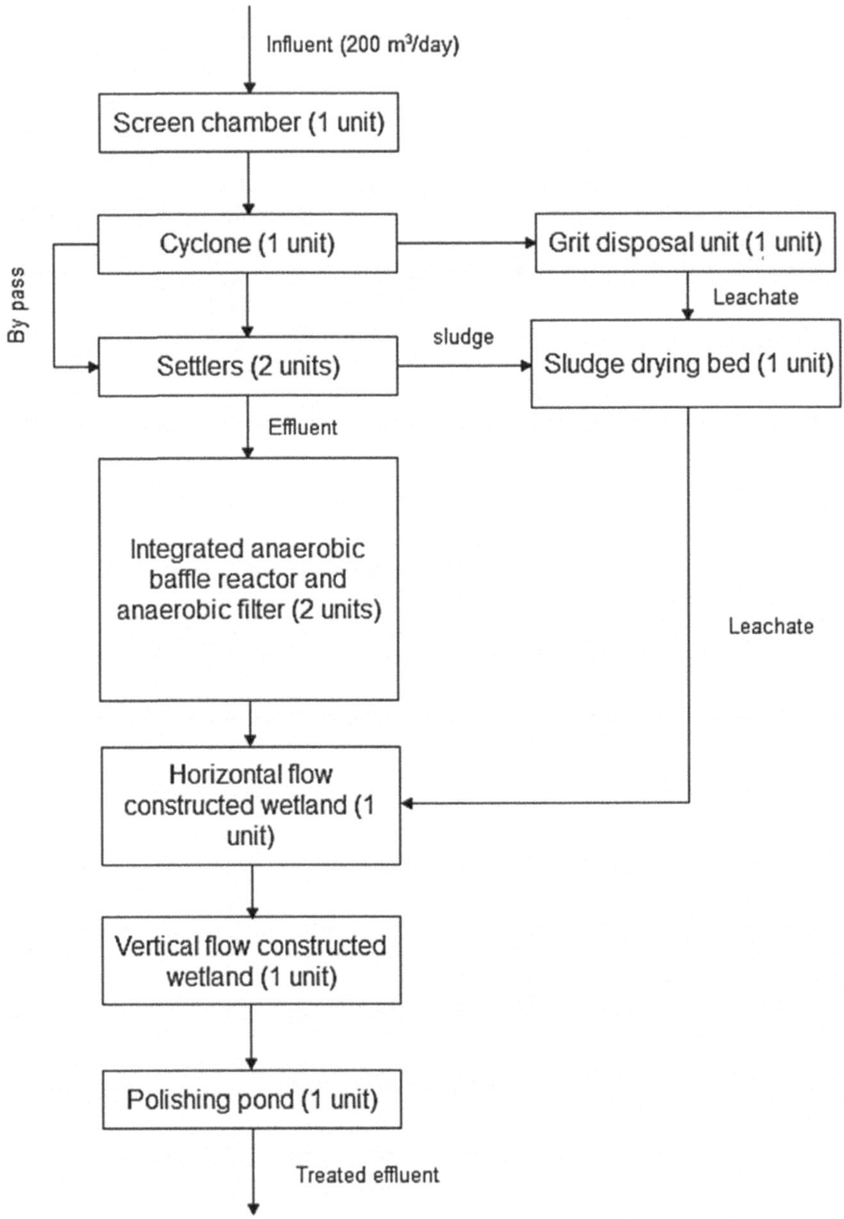

Figure 10.1 The flow diagram of the proposed full-phase Dhulikhel Hospital wastewater treatment plant.

Figure 10.2 The effluent treatment in operation at present in Dhulikhel Hospital (photo source: Krishna Ram Yendyo).

plant shows the satisfactory removal efficiency in different treatment units for the start-up operation of the constructed-wetland-based DEWATS (Figure 10.4).

It was observed that the treatment plant has satisfactory removal efficiency for TSS in all units. The performance of settlers in terms of BOD_5 removal is 34%, with COD removal of 53% and TSS removal of 71%, which meets the objective of the

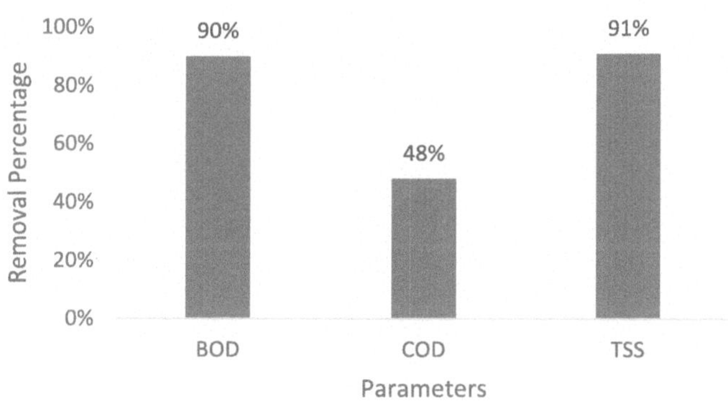

Figure 10.3 Performance efficiency of the treatment plant in 2010 (ENPHO, 2021).

Table 10.4 HWW characteristics and removal percentages in WWTP at Dhulikhel Hospital (Dugar, 2018).

S.N.	Parameters	Influent	Effluent	Removal %
1	BOD_5, mg/l	336	174	48.21%
2	COD, mg/l	1168	832	28.77%
3	Ammonia (NH_4^+), mg/l	53.5	34.5	35.51%
4	Phosphate (PO_4^-), mg/l	3.9	3.1	20.51%

design of these pre-treatment units (ENPHO, 2021). However, the BOD removal of the integrated anaerobic baffled reactor and anaerobic filter is 32%, which is less in comparison to the previous studies, which showed typically more than 60%. Since the test was conducted only a year after operation began, this might be due to the underdevelopment of sufficient anaerobic population in the biofilms or attachment media and anaerobic baffled reactor. Moreover, this study shows that the HFRB removes 21%, 14% and 32% of BOD, COD and TSS, respectively, from the effluent (ENPHO, 2021). The quality of the filter media and the type of plants used in this wetland may have affected the performance (ENPHO, 2020).

The final treated effluent has BOD_5 of 77 mg/l, COD of 175 mg/l, pH of 7.7 and TSS of 17 mg/l which meets the Nepal standard for effluent disposal from the hospital. Since the treatment plant is designed to remove mostly conventional parameters such as BOD, COD, TSS, there have been limited published results on the presence of PhACs and other emerging pollutants of concern (Dugar, 2018).

10.4.2 Sushma Memorial Hospital DEWATS

The Sushma Koirala Memorial Hospital constructed its WWTP in the year 2002 and expanded it in 2007 due to an increase in the number of beds. The designed capacity of the treatment plant is 20 m^3/d (for 250 patients). The WWTP consists

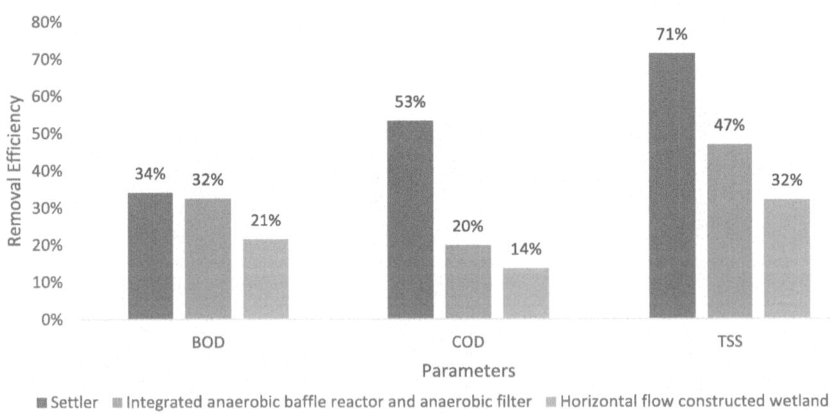

Figure 10.4 Removal efficiency of treatment units in 2021 (ENPHO, 2021).

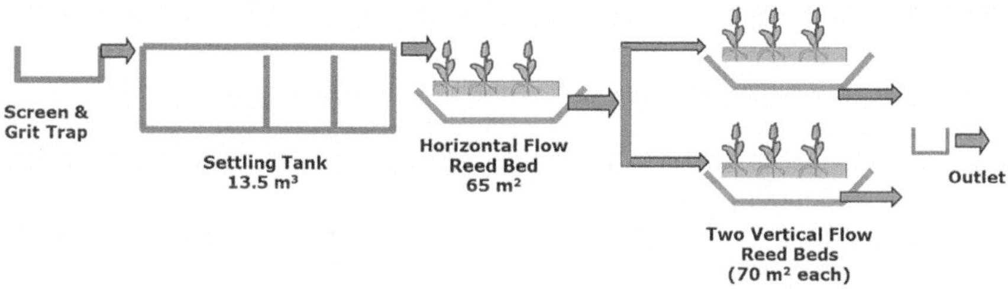

Figure 10.5 Schematic diagram of three-stage DEWATS at Sushma Koirala Memorial Hospital.

of a three-stage treatment that is, three baffled settling tanks (first stage), one HFRB unit (second stage), and two VFRB constructed wetlands (third stage). It has an inlet tank with grit cage followed by settling tank with intermittent discharge then a HFRB constructed wetland and lastly two VFRB units. The schematic diagram with the size of the units of the WWTP is shown in Figure 10.5. Figure 10.6 shows a photo of the VFCW unit of the treatment plant. The treatment plant works under a gravity flow system. The spatial footprint of the treatment plant is 200 m^2.

Table 10.5 presents the performance of the DEWATS at the hospital, and shows that the treatment performance met the existing effluent discharge standards during the year of installation (i.e., 2002), while a more recent report from Dugar (2018) shows the effluent quality does not meet the existing HWW effluent standards. The recent performance indicates that the system needs proper maintenance, as wetlands tend to become clogged during operation and need periodic cleaning and maintenance.

Figure 10.6 Vertical flow reed bed constructed wetlands at Sushma Koirala Memorial Hospital at Shankharapur Municipality, Kathmandu, Nepal (Photo: Nivesh Dugar, 2018).

Table 10.5 HWW characteristics and removal percentages in WWTP at Sushma Koirala Hospital.

S.N.	Parameters	(ENPHO, 2010)*			(Dugar, 2018)		
		Influent	Effluent	Removal %	Influent	Effluent	Removal %
1	BOD_5, mg/l	317	10	97%	330	162	50.91%
2	COD, mg/l	860	56	93%	1520	609	59.93%
3	Ammonia (NH_4^+), mg/l	–	–	–	21.8	12.4	43.12%
4	Phosphate (PO_4^-), mg/l	2.4	2.0	30%	3.1	2.4	22.58%

*Sampling was done during the year 2002.

10.5 FEASIBILITY STUDY FOR EFFLUENT TREATMENT IN TWO HOSPITALS

10.5.1 Birendra Sainik Hospital

The Birendra Sainik Hospital has a 635-bed capacity and offers a wide range of multi-speciality services to meet the medical and healthcare needs of army troops, ex-army personnel, and their relatives, as well as having a civilian facility (EcoConcern, 2019). It is a well-known hospital that has been offering quality healthcare services as well as educating enthusiastic students to pursue their goals of becoming medical professionals. It is in Chauni, Kathmandu, and spans a large area with a pleasant environment. The hospital has a population of around 1000 persons, including employees and patients. To accommodate such a large population, vital resources such as water, as well as wastewater generated by the hospital/laboratory and laundry, are used to their full potential. The release of untreated wastewater into the environment has serious consequences, especially when the wastes are chemical and contain pathogenic microbes. The army hospital has established a reputation for itself by providing excellent care and adhering to strict ethical guidelines that promote happiness and trust. Because of the critical wastewater problem, it has commenced the construction of a treatment facility.

The hospital has been utilizing 40 000 litres of water per day for laundry, which is then discharged as chemical wastewater due to its large capacity. In the hospital, there has been an issue with the generation of large amounts of wastewater and the lack of a specific treatment option for these pollutants. The feasibility study by EcoConcern recommended the installation of an electro-coagulation technology in the wastewater treatment facility to address this ongoing issue. After installation, the system can process approximately 10 KLD of laboratory wastewater and 40 KLD of laundry wastewater. This approach was proposed after studying the plant's designated region, surface-level topography, and hospital requirement assessment.

10.5.2 Norvic International Hospital

Norvic International Hospital, a medical specialist centre that was founded in 1994 by the Chaudhary Group, is a 150-bed, state-of-the-art hospital (Norvic

Hospital, 2021). The hospital's waste management is a vital source of concern. The garbage from the HCF is segregated and transferred to recyclers, particularly solid garbage. The solid waste pieces left over from the recyclers are disposed of in the municipal solid waste system, which ensures complete disinfection. The hospital is constantly concerned about the treatment of wastewater generated by its daily operations.

Wastewater from 200 beds would indeed be treated in the proposed wastewater treatment plant. The hospital is able to separate the wastewater streams to the greatest extent practicable. There is no separate pipeline for laboratory liquid waste. The wastewater at the sewer's discharge point is a mix of several wastewaters created by the hospital and stormwater from the surrounding area. Thus, the available land within the hospital ground should be evaluated in terms of its appropriateness and utility in meeting the wastewater treatment plant's goals.

The Norvic International Hospital wastewater comprises the different wastewater properties of each facility, including hospital buildings, administration buildings, laundry buildings, a canteen, and many others. Most of the buildings in the compound release domestic effluent. As a result, domestic wastewater will account for most of the wastewater volume. The idealistic manner of treating wastewater is to treat it according to its natural state, although this may not always be feasible. In this instance, various possibilities must be investigated to find the best solutions in the current circumstances. Based on our knowledge and experience, EcoConcern has suggested some options.

The feasibility study conducted by EcoConcern suggests use of a moving bed bio-film reactor (MBBR) as a potential treatment technology for treating the HWW. MBBRs can offer smaller spatial footprints and take up the variation in pollutant load and flow. In MBBRs, carrier materials are used to retain the treatment biofilms known to be more robust compared to microorganisms in the suspended growth process (Santos *et al.*, 2020). This media is kept fluidized in the aerated reactors, ensuring that the biomass receives adequate oxygen and substrate. MBBRs can be adopted for aerobic (for the removal of BOD_5, COD and ammonia) and anoxic (denitrifications) operations (Andreottola *et al.*, 2000).

However, the selection of appropriate technology for the treatment of HWW depends on the available financial resources for capital and operational cost, the area available for the WWTP, human resources for operation, and the level of treatment required for discharge (national effluent limits) or reuse (Al Aukidy *et al.*, 2018). Each practice has its strengths and weaknesses. In the case of reuse of treated wastewater, multi-barrier approaches, such as disinfection processes coupled with membrane filtration, should be adopted for better removal of harmful pathogens and other micropollutants.

10.6 PUBLIC HEALTH CONCERN

The current scenario of waste and wastewater management in hospitals presents a significant risk to public health. Only a few hospitals in Nepal treat their waste and wastewater which is a major public health concern. A study that was carried out by the United Nations Industrial Development Organization

(UNIDO) found that hospitals, clinics, and laboratories produce the most hazardous wastes that often include waste belonging to xenobiotics, nuclear material, and so on. (Adeoye *et al.*, 2018). However, the people most at risk may be the medical staff who are in contact with the wastes for a longer period, and when those wastes are discarded untreated and disposed of they could cause serious health effects (Adeoye *et al.*, 2018).

HWW contains hazardous chemicals, medicine residues like antibiotic-residues, pathogens and radioisotopes that can directly cause chemical, biological, and physical risks to the public and the environment. Most importantly the untreated wastes can result in various types of physical and pathological effects on human health (Biswal, 2013). The infected syringes and needles are also one of the important components of the hospital wastes that are often dumped into the water sources. Biomedical wastes have a greater effect on health and the environment as they can easily enter into water bodies such as surface rivers, canals and so on. The microbes that are found in the HCW can cause various diseases such as cholera, plague, hepatitis B, and diphtheria. Greater exposure to radioactive wastes can lead to nausea, vomiting, hair loss and in some people they can even lead to neurological defects.

10.7 CHALLENGES AND OPPORTUNITIES

The major challenge in hospital waste and wastewater management in Nepal is in effective implementation of guidelines and standards, and timely monitoring from the authorities concerned. Although guidelines and standards defined by the government are available, many hospitals have not implemented them effectively. One of the reasons for this could be inadequate budget allocation due to less prioritization of issues related to waste and wastewater management in hospitals, the lack of management practices in the hospital to implement the treatment plants, and insignificant monitoring from the government sector.

The recent COVID-19 situation has highlighted the importance of proper waste and wastewater management to protect public health due to which the interest of the hospital management in waste and wastewater treatment has increased. Although the low-cost decentralized wastewater treatment systems, such as wetlands, are common in HWW management, hospitals have shown interest in more advanced technologies, such as electro-coagulation technology and MBBRs, in recent years.

10.8 CONCLUSION AND RECOMMENDATIONS

Some past and recent initiatives regarding HCW and HWW management practices in hospitals can be seen in Nepal. Most of the hospitals have prioritized management of solid waste more than wastewater. Although a few hospitals are effectively managing both their waste and wastewater, more initiatives are essential to avoid contamination and reduce public health risks in Nepal. The multitude of individuals involved, as well as combined efforts from the government/public sectors and also the private and non-governmental sectors, are expected to improve the situation. In addition, various treatment technologies including those from small-scale private initiatives have also helped to improve the overall HCWM. It is vital to conduct more research on improving the existing treatment systems and ensuring their continuous monitoring and support from the agencies concerned.

AUTHORS' CONTRIBUTIONS

Conceptualization-Coordination-Compilation, A. Gautam; Content Design, A. Gautam., K. R. Yendyo, and P. Shrestha, A. Ghimire; Case Studies Presentation-Formal Analysis, P. Shrestha., I. Manandhar, A. Ghimire, and K. R. Yendyo; Writing-Editing-Formatting, P. Shrestha., A. Gautam., S. Kunwar, and I. Manandhar; Review-Editing-Networking, S. Panthi, S. Kunwar, A. Ghimire, K. R. Yendyo. All authors have read and agreed to the published version of the manuscript.

ACKNOWLEDGEMENTS

The authors are grateful for the support and suggestions from Rajendra Shrestha representing the Environment and Public Health Organization (ENPHO), Nepal; Rajiv Joshi and Mahesh Neupane from the Department of Water Supply and Sewerage Management (DWSSM), Nepal; and the Society of Public Health Organization (SOPHEN), Nepal. Special thanks to Dr. Roshan Raj Shrestha – Deputy Director at Bill & Melinda Gates Foundation and Waterlution/Global Water Innovation Lab 2020 Team for the consultation and networking opportunities.

REFERENCES

Al Aukidy M., Al Chalabi S. and Verlicchi P. (2018). Hospital wastewater treatments adopted in Asia, Africa, and Australia. In: Hospital Wastewaters: Characteristics, Management, Treatment and Environmental Risks, P. Verlicchi (ed.), Springer International Publishing, Cham, Switzerland, pp. 171–188, https://doi.org/10.1007/698_2017_5

Andreottola G., Foladori P., Ragazzi M. and Tatàno F. (2000). Experimental comparison between MBBR and activated sludge system for the treatment of municipal wastewater. *Water Science and Technology*, **41**(4–5), 375–382, https://doi.org/10.2166/wst.2000.0469

Adeoye AO., Akadnde E. A. and Lateef A. (2018). Impacts of hospital waste management on the health and environment of ogbomoso area, Oyo state. *Hospice and Palliative Medicine International Journal*, **2**(6), 386–389, https://doi.org/10.15406/hpmij.2018.02.00130

Asian Development Bank. (2013). Initial Environment Examination Report of Kathmandu Valley Wastewater Management Project Kathmandu.

Biswal S. (2013). Liquid biomedical waste management: An emerging concern for physicians. *Muller Journal of Medical Sciences and Research*, **4**(2), 99, https://doi.org/10.4103/0975-9727.118238

CWISTA Hub. (2019). Fecal Sludge Management for Integration into Kathmandu Valley Wastewater Master Plan.

DoHS. (2019). Annual Report: Department of Health Services 2075/76 (2019/19). Department of Health Services, Ministry of Health and Population, Government of Nepal, **76**, 2–2, https://publichealthupdate.com/department-of-health-services-dohs-annual-report-2075-76-2018-19/ (accessed on September, 8, 2021)

Dugar N. (2018). Study of Fate of Pharmaceutically Active Compounds (PhACs) in waste water and their treatment plant in Nepal. MTech thesis, Department of Environmental Science and Engineering. Kathmandu University, Dhulikhel, Nepal.

EcoConcern. (2019). Health Care Waste Management in Hospitals in Kathmandu, Current Situation and Role of Private Sector.

ENPHO. (2010). DEWATS For Sushma Koirala Hospital.

ENPHO. (2015). Factsheet on Decentralized Waste Water Treatment System (DEWATS) for Dhulikhel hospital.

ENPHO. (2017). Establishment of Decentralized Waste Water Treatment System (DEWARS) in Dhulikhel Hospital Final Report.

ENPHO. (2020). ENPHO Technical Field Visit in Dhulikhel DEWATS. Environment and Public Health Organization.

ENPHO. (2021). ENPHO Lab Report. Environment and Public Health Organization.

Green H., Poh S.-C. and Richards A. (2003). Wastewater Treatment in Kathmandu, Nepal. 1–25, http://web.mit.edu/watsan/Docs/Student Reports/Nepal/NepalGroupReport 2003-Wastewater.pdf (accessed on September, 8, 2021)

HCWM SOP. (2020). National Health Care Waste Management. Standards and Operating Procedures – 2020. Government of Nepal. Ministry of Health and Population.

Jha A. K. and Bajracharya T. R. (1998). Wastewater treatment technologies. *Metal Finishing*, **96**(6), 148, https://doi.org/10.1016/s0026-0576(98)80321-8

Joshi H. D., Acharya T., Ayer R., Dhakal P., Karki K. B. and Dhimal M. (2017). Health care waste management practice in health care institutions of Nepal. *Journal of Nepal Health Research Council*, **15**(35), 7–11, https://doi.org/10.3126/jnhrc.v15i1.18006

MoUD. (2014). National Water Supply and Sanitation Sector Policy 2014. Ministry of Urban Development (MoUD), Government of Nepal. Retrieved from: https://dwssm.gov.np/wp-content/uploads/2019/05/WSS-Policy-Dec14.pdf (accessed September, 9, 2021)

MOHP. (2014). Health Care Waste Management Guideline Health Care Waste Management Guideline Ministry of Health and Population.

MOHP. (2020). Health Care Waste Management in the context of COVID-19 Emergency. June.

Norvic Hospital. (2021). Norvoc Internaationaal Hospital Healthcare Medical Tourism.

Pathak D. R., Nepal S., Thapa T., Dhakal N., Tiwari P. and Sinha T. K. (2021). Capacity assessment and implementation analysis of common treatment facility for the management of infectious healthcare waste in rapidly urbanising city of Nepal. *Waste Management & Research*, **39**(1_suppl), 64–75, https://doi.org/10.1177/0734242X211013910

Rajendra R. B., Robert D. and Loannis K. (2016). Historical Development of Wastewater Management in Kathmandu Valley.

Santos A. D., Martins R. C., Quinta-Ferreira R. M. and Castro L. M. (2020). Moving bed biofilm reactor (MBBR) for dairy wastewater treatment. *Energy Reports*, **6**, 340–344, https://doi.org/10.1016/j.egyr.2020.11.158

Song Y., Zhang M., Yin L., Wang K., Zhou Y., Zhou M. and Lu Y. (2020). COVID-19 treatment: close to a cure? A rapid review of pharmacotherapies for the novel coronavirus (SARS-CoV-2). *International Journal of Antimicrobial Agents*, **56**(2), 106080.

Verlicchi P., Galletti A., Petrovic M. and Barceló D. (2010). Hospital effluents as a source of emerging pollutants: an overview of micropollutants and sustainable treatment options. *Journal of Hydrology*, **389**(3), 416–428, https://doi.org/10.1016/j.jhydrol.2010.06.005

WHO. (2005). Management of Solid Health-Care Waste at Primary Health-Care Centres A Decision-Making Guide Management of Solid Health-Care Waste at Primary Health-Care Centres. World Health Organization.

WHO. (2020). Water, sanitation, hygiene, and waste management for the COVID-19 virus. Interim Guidance. World Health Organization.

doi: 10.2166/9781789062625_0203

Interview Section

Based on Current Treatment Processes and Future Prospects w.r.t Industrial Points of View Provided by Niloofar Mozaffari & Nastaran Mozaffari

Interviewees: Rowe Michels, Rick McGregor, Shannon Spurlock

INTERVIEW 1

1. **Please introduce yourself, your company, and what activity you are doing/have done in the water sector.**
   Rowe Michels is the Chief Executive Officer of Alpheus Water Research, a consultancy and fund that is entirely focused on new, disruptive and environmentally friendly technologies to generate or reuse water. Previously, Rowe was a Wall Street Executive and Research Analyst, gaining 11 consecutive #1 rankings for his coverage of the Electric and Water Utilities Sector.
2. **What is your perspective in hospital wastewater treatment?**
   Similar to all wastewater, the safe and cost-effective management of hospital wastewater is increasingly challenged by ever more types of chemicals including many new pharmaceuticals and chemical treatments. These new 'emerging contaminants' are particularly challenging as they have very long lives (i.e., 'forever chemicals') and/or are nano-sized, making traditional filtration difficult or ineffective. Hence, the prior 'dilution solution' (i.e., which assumes only small amounts become irrelevant in vast water bodies such as oceans), which has been an assumption for the wastewater sector for decades, is becoming irrelevant as even trace amounts of very small 'forever chemicals' accumulate over time. The sector will need new technologies to better separate out various contaminants and toxins so that they can be reused, or fully destroyed.
3. **What challenges are you looking for in this water operator position?**
   N/A

4. **What type of future can we expect from hospital wastewater treatment plants if they existed?**
   Hospital wastewater effluents will increasingly need to be treated locally, avoiding excessive transportation costs, and with a greater focus on recapturing, recycling and reusing certain high value chemicals and elements. In addition to transportation costs, most filtration or thermal heating (to evaporate) systems require very high amounts of energy, which implies more cost and C02. Thus, new treatment systems will also ideally use less energy, and also less chemicals as these create another layer of environmental impact.

5. **What are new cost effective systems that can replace existing treatment systems?**
   New water treatment systems will be smaller and more distributed, allowing for less transportation/storage costs and more recapture, recycle and reuse. Ideally, they will require less energy and chemicals, but will be better able to separate out certain target chemicals for reuse or customized disposal. Some of the most promising new technologies that meet these goals include forward osmosis, biomimicry-based enhanced evaporation, custom developed microbes, lower-energy ionization, and conventional disposal (such as reverse osmosis or crystallization/ZLD) that is powered by renewable clean energy.

INTERVIEW 2

1. **Please introduce yourself, your company, and what activity you are doing/have done in the water sector.**
   Rick McGregor, I am senior technical advisor of two companies, InSitu Remediation Services Ltd and HClearO both based in Canada but do work around the world. IRSL specializes in the in situ treatment of groundwater using a variety of chemical, physical and biological approaches whereas HClearO focuses on the above ground treatment of impacted groundwater using a mix of chemical, physical and biological processes.

2. **What is your perspective in hospital wastewater treatment?**
   For us, hospital wastewater has two main areas of concern, traditional contaminants such as organic carbon, volatile organic compounds, heavy metals, petroleum hydrocarbons, etc. and emerging contaminants such as pharmaceuticals, PFAS, micro plastics, viruses etc. For groundwater impacts the treatment of the first set of compounds are relatively well understood and demonstrated using both in situ and ex situ methods. For the emerging contaminants, ex situ methods are better demonstrated and advanced than in situ methods which are still being researched, tested, and implemented. However, the in situ treatment of these compounds will most likely be more sustainable and economically more appealing than ex situ methods.

3. **What challenges are you looking for in this water operator position?**
Water operator positions will continue to demand higher education and training to operate the systems as they become more complex and automated. This combined with the drive towards flexible systems that can treat highly specific compounds will require the operators to have a more diverse educated/experience background and be able to problem solve or require more reliance on specialized engineering/maintenance services.

4. **What type of future can we expect from hospital wastewater treatment plants if they existed?**
The complexity of wastewater derived from hospitals will continue to increase as more compounds are introduced into the medical system. This combined with increased understanding of human and ecological toxicology and greater demand for potable water and usable land will demand that treatment plants/systems become more efficient and effective.

5. **What are new cost effective systems that can replace existing treatment systems?**
A drive away from chemical/mechanical/energy intensive systems will continue with biological/natural based systems becoming more common. This change will be challenged by the need to treat more compounds within the wastewater to lower levels.

INTERVIEW 3

1. **Please introduce yourself, your company, and what activity you are doing/have done in the water sector.**
My name is Shannon Spurlock and I started my own consulting firm, Ochotona LLC, at the beginning of the COVID-19 pandemic, in March 2020. Prior to this endeavor, I worked in a role where I advanced equitable food policy and built strategic partnerships; it was in this role that I began working to expand the allowed applications of recycled water in Colorado (USA). Specifically, I built support across to advance policies that would allow Colorado growers to be able to use recycled water to irrigate crops grown for human consumption. It became very clear that by recycling water, we would be able to maximize a limited resource and create a new water supply for agriculture that was safe, reliable, and drought resistant.

2. **What is your perspective in hospital wastewater treatment?**
Treating wastewater to appropriate standards is important to human health and the environment. By treating hospital wastewater, there is an opportunity to remove contaminants and reuse the water safely. This is important as it can provide a new water source that is safe, reliable, and drought resistant. Additionally, testing wastewater is an effective method of detecting diseases and has been used in the past to detect polio and currently, is being used to understand COVID-19 rates within a community.

3. **What challenges are you looking for in this water operator position?**
   N/A

4. **What type of future can we expect from hospital wastewater treatment plants if they existed?**
   Though I don't work in the hospital system, it is my belief that having onsite wastewater treatment plants would allow hospitals to treat wastewater before releasing it into the environment. This has many benefits – ranging from ensuring that the water is safe and free from harmful chemicals and contaminants to creating a localized water resource that can be treated and appropriately reused.

5. **What are new cost effective systems that can replace existing treatment systems?**
   (I'm not able to answer this question as I don't work on the operator/treatment side of recycled water.)

Index

E

F

G

H

I

M

N

O

P

R

S

T

U

V

W

IWA Publishing's authorised EU representative for General Product Safety Regulations is Diane D'Arras, 15 rue Duret, 75116 Paris, France, e-mail: safety@iwap.co.uk.

Printed and bound by CPI Group (UK) Ltd, Croydon, CR0 4YY
16/06/2026
02140885-0001